AF291164

Management, Change, Strategy and Positive Leadership

Series Editors

Satinder Dhiman
School of Business
Woodbury University
Burbank, CA, USA

Joan Marques
School of Business
Woodbury University
Burbank, CA, USA

Preface

...Technology is propelling social networks and cybertools [sic] deeper and deeper into our lives, our privacy and our politics—and democratizing the tools for 'deep fakes,' so that many more people can erode truth and trust. But the gap between the speed at which these technologies are going deep and the ability of our analog politics to develop the rules, norms, and laws to govern them is getting wider, not narrower. That gap has to be closed to preserve democracy.[1]
Thomas L. Friedman, *The New York Times*

According to Casper Klynge, who represents Denmark's interests as an ambassador to the big tech companies in Silicon Valley, "[Companies such as Facebook and Google] have moved from being companies with commercial interests to actually becoming de facto foreign policy actors."[2]
Adam Satariano, *The New York Times*

The tech industry failed to anticipate [the intense backlash against it] 'because the nerds had ascended, culturally and socially, and had become enchanted with their own virtuous self-image'...Rather than abandon the tech industry, we should 'build it anew with fore-thought.'[3]
Nicholas Thompson, *Wired*

Countless books already exist that sing the unadulterated virtues of technology. This is not one of them. It is a hard-hitting, no holds-barred examination of the dark sides of technology.

The authors have spent their careers helping individuals and organizations overcome Denial, so that they could address the truly difficult problems and potential crises facing them. It's time for society as a whole to do the same. We need to face head-on the perils of modern technology.

While both the content and tone of the book about the current state of technology will undoubtedly be pessimistic to many, our ultimate message is nonetheless one of hope. We remain optimistic about not only what is needed, but can be done to make technology serve the greater good.

---

[1] Thomas L. Friedman, "The biggest Threat to America Is Us," *The New York Times*, Wednesday July 3, 2019, p. A23.

[2] Adam Satariano, "When Tech Firms Are Superpowers, Send an Ambassador," *The New York Times*, Wednesday, September 4, 2019, p. B1.

[3] Nicholas Thompson, "Rebuilding Big Tech the right way," *The Week*, October 25, 2019, p. 38.

Given that both authors have doctorates in engineering, we are the last to deny the countless benefits of technology. Instead, we need drastically different and better ways of managing it not just for our benefit, but even more for future generations.

In brief, we are staunch proponents of the urgent need for Socially Responsible Tech Companies. For this reason, we lay out a specific proposal for the creation of a new government office for socially responsible technology or OSRT. Its main purpose is to help ensure that before any technology is released and over its entire lifetime that it passes the most stringent tests so that it does indeed serve the public good, and least of all, does not do irreparable harm.

Along with the need for substantial change in the reward systems of tech companies, the education of technologists needs to change significantly as well. Ethics needs to be a fundamental part of the thinking of technologists. Technical skills alone are no longer sufficient. Running tech companies for the greater good necessitates people and organizational skills as well. More than ever, it also requires the ability to consider the societal impacts of technology.

True, many companies already have chief privacy and technology officers, but they do not necessarily work together, let alone seamlessly.[4] The same is true of their working together with the chief strategy and crisis management officers of an organization. In short, both the management and the leadership of tech firms need to be systemic or it won't work at all.

> In a word, technology, which has made our lives incomparably better—for one, we live longer and healthier lives-now constitutes one of the biggest threats facing humankind. One of the crowning achievements of Western Civilization, and long regarded as the saving grace of humankind, technology has instead become a scourge. It's nothing less than an existential threat of the gravest order.

The Historic Role

For most of human history, technology has played a largely limited and thus greatly confined role. Even though it's constantly threatened to remake and/or to replace us either in whole or in part, and thus has continually muddied the boundaries between humans and machines, it's served primarily to augment our limited physical and mental abilities. But more recently, it's taken on a greater and much more highly worrisome role. It threatens to eradicate once and for all the heretofore sacrosanct boundaries between humans and machines. We possess as never before the godlike powers to intervene and thus alter every aspect of the human condition, the most

---

[4]See Kenneth A. Bamberger and Deidre K. Mulligan, *Privacy on the Ground: Driving Corporate Behavior in the United States and Europe*, MIT Press, Cambridge, Mass, 2015.

troublesome being our genetic makeup.[5] The supreme, if not ominous, question facing us is, "Who and what is, and even more, will be human in the not-so-distant future?" To our great misfortune, if not eventual undoing, we lack the accompanying godlike wisdom to know when *not* to employ such God-altering technologies just because we possess them.

Thinking the Unthinkable

This book is fundamentally about the wisdom necessary to manage technology for the betterment of humankind. It requires doing one of the most difficult tasks with which humans are ever tasked: thinking long and hard about what can go terribly wrong as the result of our godlike powers and then doing everything possible to prevent it from happening. It demands doing nothing less than Thinking the Unthinkable.

Increasingly, in our quest to become the masters of human evolution, we're playing with forces of whose ultimate consequences we have little knowledge, and as a result, even less control.[6] With the best of intentions, virtually all of the marvelous inventions that are supposed to make our lives dramatically better have, with few exceptions, produced strikingly opposite effects. More and more crises are thereby the norm. Every crisis is in fact "Unthinkable" until it occurs. If we have not already reached it, we are soon approaching the point where the disbenefits of technology greatly outweigh its benefits.

Social Media and the Internet are prominent examples. Instead of as they promised bringing us closer together and making us better informed, they have served as highly efficient vehicles for the spread of dis- and misinformation, not to mention hate speech and out-and-out conspiracies, thereby driving us further apart. Indeed, when confronted with the fact that it has carried political ads that are full of egregious lies, Facebook has steadfastly refused to remove them thus bowing to the enormous pressures of right-wing conservative groups that it favors liberal points of view.[7] It's hidden behind the spurious claim that it's acting in the "best interests of the freedom of speech." Or, "politicians lie all the time so why should we be held responsible for holding them to account?" This of course flies directly in the face of the fact that they are directly responsible for providing one of the most powerful vehicles—platforms—ever invented by which to spread lies.

---

[5]See Jon Cohen, "Chin's CRISPER Revolution," *Science*, August 2, 2019, pp. 420–421; Jennifer A. Doudna and Samuel H. Sternberg, A Crack In Creation: Gene Editing and the Unthinkable Power to Control Evolution, *Mariner Books*, New York, 2018; Samuel Tawfick and Yichao Tang, "Stronger artificial muscles, with a twist," *Science*, July 12, 2019, pp. 125–126.
[6]*Op cit*, Doudna and Sternberg.
[7]Cecilia Kang, "Facebook Won't Pull Ads That Lie," *The New York Times*, Wednesday, October 9, 2019, p. B1.

It also flies in the face of the fact that unlike the government, private organizations are not obligated to present, let alone defend, all points of view. They are free to operate by means of their stated values. To be sure, it would be foolish to expect everyone to agree with their policies, but then strong disagreements and opposition come with the basic territory.

As an editorial in the *San Francisco Chronicle* put it:

> Either Facebook claims both the rights and responsibilities of being an information platform with a global reach, or it purposefully allows truth-denying discourse to warp the environment that enabled it to become successful. Zuckerberg seems to be betting that the later won't damage Facebook, even if it damages democracy.[8]

In short, Social Media and the Internet magnify our worst impulses. One of the worst examples by far is the fact that by making it far too easy for young, vulnerable minds to be exposed and consequently become unduly influenced by white nationalist propaganda, Social Media have thereby played a major role in the spread of Domestic Terrorism.

No less damaging is the abject failure of tech companies to take down sexually explicit photographs and videos of young children even though they have the means to do it.[9] It's proven nothing less than a boon for pedophiles and the like. The harm done lasts for years, if not a lifetime.

This book is not only a systematic guide to Thinking the Unthinkable, but the strongest argument we can muster for the proposition that before any technology is released, and over its entire lifetime, tech companies and technologists are not only obligated but need to be required to plan for the Unthinkable, and thus, the inevitable crises, big and small that they engender. They certainly need to pay serious attention to the disbenefits that accompany everything that humans create and do. Most important of all, they need to do everything in their power to thwart the Unthinkable before it happens.

A strong word of caution is in order. There is no single sure-fire way for Thinking the Unthinkable. Anyone who is looking for a single scheme that will positively guarantee success will be greatly disappointed. Instead, we offer a series of interrelated methods that are the best of which we know for engaging with a difficult topic of the greatest importance. It requires continuous critical thinking of the highest order.

---

[8]Editorial, "Facebook's distortion defense," *San Francisco Chronicle*, Wednesday, October 23, 2019, p. A10.

[9]Michael H. Keller and Gabriel J. X. Dance, "Child Sex Abuser Elude Flimsy Digital Safeguards," *The New York Times*, Sunday, November 10, 2019, p. 1.

Systems Thinking and Crisis Management (CM)

It's our fervent hope that one of the most important takeaways of the book is how to think systematically about the innumerable crises that accompany all technologies so that we can do a far better job of anticipating and thereby preventing them before they happen. To help accomplish it, as we proceed, we review a limited number of crises for the critical lessons they have to teach. Nonetheless, we cannot emphasize enough that this is not a book about CM per se. Instead, it's about the bare essentials of CM that every citizen needs to know so that we can ensure that tech companies are operating in our best interests, indeed to hold them better accountable.

Reactive CM Is the Norm

A front-page article in *The New York Times* on Amazon provides both a necessary and telling example.[10] Amazon is responsible for at least half of all the book sales in the USA. This not only allows it to crush the competition, but gives it the unrivaled power to force the publishing industry to agree to the prices and terms it sets for books. If a particular publisher doesn't go along with its demands, then it refuses to carry its books.

Worst of all, Amazon has not only carried counterfeit books, but done virtually nothing to curtail their both proliferation and dissemination. One of the most disturbing cases is that of a phony medical book whose print is so blurry that it's literally indecipherable. As a result, the correct dosages of the medications doctors are supposed to give are so unclear that it puts the health of patients seriously at risk.

No less disturbing is Amazon's attitude and behavior toward CM. Indeed, it's reflective of the tech's industry general reactive stance. In brief, "Don't do anything meaningful until the crises are so overwhelming that we're finally forced to act. And, then just 'stonewall it' for as long as possible by giving one meaningless apology or excuse after another." In this way, the lack of any acceptable response only makes the initial crises worse.

For this reason, in 2014, UK Baroness, Beeban Kidron started the foundation 5Rights to promote children's digital rights because of her frustrations with the major tech companies. "They scrambled to handle one problem after another for children on their sites, [Beeban] said, but appeared unwilling to make major changes to try to avert such problems in the first place. She concluded that the only way to give children more privacy and more control over their online experiences was through regulation."[11] We couldn't agree more.

---

[10]David Streitfeld, "Amazon's Control Over Books Shows the Peril of Tech Power," *The New York Times*, Monday, June 24, 2019, pp. A1–A12.

[11]Singer, *op cit*, p. B6.

In fact, when it comes to protecting the rights of children, the European Union or EU is far ahead of the USA. In fact, the EU has a "Digital Czar" who appropriately is regarded as the "Most Powerful Regulator of Big Tech."[12]

Even more, the EU has recently taken the unprecedented step of ruling that individual member countries can not only take down Facebook posts, photographs, and videos that it finds inflammatory, but as a result, so can other countries far beyond their immediate borders.[13]

This is not to say that tech companies are doing nothing at all. For instance, we know of a company that has a "Chaos Team" that is charged with investigating "surprises," i.e., cases where "unexpected incidents" have affected its software and thus disrupted its operations. And it does it without blaming those involved. Regrettably, it does not look for "surprises" in advance, i.e., proactively. It also doesn't coordinate with other units that are on the lookout for malicious actors. As such, it does not practice Proactive Crisis Management, the essence of which is thinking and behaving systemically, and thus working with all the various units involved in CM. Once again, this appears to be the general norm of tech companies. For this reason, say more about the nature of CM in a later chapter.

Exemplary Behavior

The contrast between the behavior of tech companies and that of other industries couldn't be greater. Thus, after the 1982 poisonings of Tylenol capsules in a suburb outside of Chicago, McNeil Pharmaceuticals, makers of Tylenol, didn't hesitate to withdraw all the bottles of its painkillers off the shelves nationwide until they could ensure the safety of consumers. In this way, they restored trust in the company, if not in the entire industry. In fact, the exemplary reaction of McNeil Pharmaceuticals to the Tylenol poisonings was the impetus for the founding of the modern field of CM of which Mitroff was one of the first and principal contributors. Nonetheless, the recent exorbitant raising of prices for lifesaving drugs such as insulin shows that it's not enough to do the right thing once. All of the "good" one has done in the past can be lost in a moment. It's nothing less than a strike against the entire industry.

Recently, over 500 million judgment has been brought against Johnson and Johnson, the parent company of McNeil Pharmaceuticals, for its unsavory role in the opioid crisis. We cannot emphasize enough that all the good one has done in the past is no absolute insurance that one will continue to do good and thus benefit from it in the future. This is unfortunately one of the key lessons of CM that far too many organizations fail to learn. One has to do "the right things" everyday.

---

[12]Matina Stevis-Grindneff, "E.U.'s New Digital Czar Expnads Her Portfolio," "Most Powerful Regulator of Big Tech," *The New York Times*, Wednesday, September 11, 2019, p. A12.

[13]Adam Satariano, "E.U. Ruling on Facebook Crosses Borders," *The New York Times*, Friday, October 4, 2019, p. B1.

Proactive CM Must Be Mandated

Yes, tech is a different industry. Nonetheless, given the huge role that technology plays in Pharma, there are strong parallels with tech companies. The overriding point is that Proactive CM must be a basic part of every company's entire operations. Unless one gets out in front by preparing for a whole slate of crises before they happen, it has been shown repeatedly that one will not fare well in the heat of an actual crisis. Indeed, it only makes things worse.

While not perfect, Proactive CM is absolutely vital in anticipating the worst and doing everything we can to prevent it from happening. Furthermore, it's so critical that Proactive CM must be mandated. It cannot be left up to the discretion of tech companies to do if they feel like it. Before they're allowed to operate and over their entire lifetimes, the burden is squarely on them to show that Proactive CM is a fundamental part of their thinking and entire operations.

Future Generations

To reiterate, even though both authors have doctorates in engineering, we are not unabashed advocates of technology. This does not mean that we are staunch opponents of it. Rather, we are highly critical of the lackadaisical attitudes of far too many technologists. The lack of concern with the social consequences and impacts of technology is seriously disturbing to put it mildly, for the consequences, both intended and unintended, as well as the social impacts of technology are greater than ever.

It's perfectly true that in every way possible, our world would literally collapse without modern technology. But this means that we owe it not only to ourselves, but to future generations to ensure that we are doing everything that we can to safeguard us from its worst aspects so that they will continue to benefit from technology as we have.

Finally, we are the first to admit that this book presents an ideal about how to think and operate technology for the greater good. For this and other reasons, we don't expect tech companies to adopt our precepts anywhere soon. Nonetheless, our ardent hope is that by presenting a clear portrait, we have shed important light regarding what's needed to be done. To be perfectly frank, we believe that tech companies will have to be compelled to act in society's best interests.

Berkeley, USA   Ian I. Mitroff
Kensington, USA   Rune Storesund

Acknowledgements We wish to acknowledge the helpful and encouraging advice of Dr. Deborah Siegel throughout every phase of the book. We also wish to acknowledge Ms. Brook Warner who commented on earlier drafts of the book. Whatever flaws remain is ours.

We also wish to thank Nitza Jones-Sepulveda, our editor at Springer, for guiding the book through the process of publication.

Praise for *Techlash*

"The focus here is technology, but this analysis is relevant to all of us trying to insure that our work ends up making the world a better place not a worse place. Mitroff and Storesund show how the unintended consequence is still very much our problem and that these consequences can be anticipated and controlled."
—David Brancaccio, *Public Radio and Television Journalist*

"What a thoughtful person likes when reading is an experience that makes their thinking more incisive and trenchant. This is the pleasure and value of reading Ian Mitroff's writings. He continues to provide this important value as we adjust to how emerging technology is changing our lives."
—Gerald Harris, *President, Quantum Planning Group, Inc. (www.artofquantumplanning.com), Leader, the Technology and Society Forum, California, USA*

"The authors of this timely book have not sought to produce another Jeremiad against the risks and dangers of technology. Instead they invite us to 'think the unthinkable', that is seeking to control, tame and humanize current technologies by examining their practical and ethical implications not in isolation but in combination with each other. Thinking the unthinkable means working against psychological and social defences that create collective moral and social blindness. It also means working tirelessly on bolstering social institutions aimed at regulating and containing technological forces. Individuals, whether as citizens, consumers or employees can only have modest impact on the major challenges our societies face today. Only through collective action and innovative political institutions can these challenges be addressed."
—Yiannis Gabriel, *Professor Emeritus, Bath University, UK*

Contents

About the Authors

Ian I. Mitroff is generally regarded as one of the principal founders of the modern field of Crisis Management, about which he has written extensively.

He has a BS in engineering physics, an MS in structural engineering, and a Ph.D. in industrial engineering with a minor in the philosophy of social systems science, all from the University of California, Berkeley.

He is Professor Emeritus from USC where he was Harold Quinton Distinguished Professor of business policy in the Marshall School of Business. He also had a joint appointment in the Annenberg School for Communication.

He is currently Senior Research Affiliate in the Center for Catastrophic Risk Management at UC Berkeley.

He is Fellow of the American Psychological Association, the American Association for the Advancement of Science, and the American Academy of Management.

He has an honorary Ph.D. from the University of Stockholm and a gold medal from the UK Systems Society for his lifelong contributions to Systems Thinking.

Rune Storesund has a BA in anthropology from UC Santa Cruz and a BS in civil engineering, part of a joint "dual degree" program between the two schools which intended to "diversity" engineering students. He has a master of civil engineering and doctorate of civil engineering from UC Berkeley.

He is Professional Geotechnical and Civil Engineer, Board-Certified Diplomat of the National Academy of Forensic Engineering, and Fellow of the American Society of Civil Engineers.

He is CEO and President of his boutique engineering and risk consulting firm Storesund Consulting, CEO and President of his civil work construction company Storesund Construction Services, Inc, as well as CEO and President of NextGen Mapping, Inc., an innovative software development company.

He is Executive Director of the Center for Catastrophic Risk Management at UC Berkeley specializing in the state-of-the-art research and application of safety and reliability of critical infrastructure systems.

Threats of Imaginable and Unimaginable Proportions

> *Technology helps facilitate our relationships with people far away, but not [with] the people sitting next to us.*
> *Broder-Sumerson* [1]

Anthony Levandndoski, a former Google engineer, "…founded a church called, the Way of the Future, that was devoted to 'the realization, acceptance, and worship of a God-head based on Artificial Intelligence [AI]'. 'Machines would eventually become more powerful than humans, he proclaimed, and the members of his church would prepare themselves, intellectually and spiritually, for that momentous transition…' I don't believe in God, 'he [said]'. 'But I do believe that we are creating something that basically, as far as we're concerned, will be like God to us.'"[1]

This book is about the dire existential threat posed by modern technology. Indeed, it grows with every passing day. While much of our criticism is focused on Facebook, it's only because it's the most persistent and visible representative of what's wrong with tech in general, which is our real concern.

Fundamentally, it's about what needs to be done to make both technologists and tech companies socially responsible. If they don't change and behave more responsibly—more ethically–they will be subjected to ever more burdensome rules and regulations, thereby making it harder for us all to reap the positive benefits of technology. In fact, since technologists and tech companies have demonstrated repeatedly that they would rather fight than accept legitimate and reasonable regulations, nothing substantial will happen without the imposition of stiff penalties and clear constraints. In short, they've brought onerous rules and regulations on themselves. What these need to be—even more important, how they should be administered—is one of the major topics of the book. For this reason, we inten-

---

[1]Duhigg [2].

I. I. Mitroff and R. Storesund, *Techlash*, Management, Change, Strategy and Positive Leadership, https://doi.org/10.1007/978-3-030-43279-9_1

tionally outline a process for helping to ensure that technology will act on our behalf. By themselves, all the rules and regulations will not do the job.

A backlash against technology—Techlash!—has been brewing steadily. It's due to the confluence of a number of important factors. By itself, each is powerful enough, but acting together, their impact is magnified considerably. The following is merely a sample of the many issues we treat throughout:

1. The monopolistic, predatory power of tech companies and thus the growing chorus to break them up; in a word, they are the "new robber barons";
2. The shameful, out-and-out unethical behavior of tech companies, most notably Facebook, not to mention Instagram, Google,[2] etc., that continue to come to light, virtually on a daily basis; they violate not only our privacy but our deepest sense of self and well-being by collecting without our explicit knowledge and consent enormous amounts of personal data which they then sell to unscrupulous third parties for their profit;
3. The fact that technology and technologists always promise more than they can possibly deliver; in a word, they overly hype the positive benefits of their creations with little if any thought with regard to their inevitable negative and unintended consequences; much is due to serious shortcomings in the education of technologists—namely the lack of serious programs in ethics and management; as such, and not always for the better, technology affects society as a whole;
4. And not least of all, the fact that those democratic and republican members of Congress who were once staunch supporters of tech companies such as Facebook and Google—in large part, because they contributed substantially to their campaigns, and thereby let them operate—if not get away—with few if any regulations at all are so now angry at their outrageous behavior that they are now calling for tough regulations.[3] It even brought democrats and republicans together in one of the few bipartisan issues on which they can both agree.

Not surprisingly, members of Congress have been focused primarily on the monopolistic, predatory power of tech companies. While clearly important, they have given less attention to the negative aspects and unintended consequences of technology. Nonetheless, this is changing as well.

---

[2]See Singer and Conger [3].
[3]Lohr [4].

Dangers

Even though it was exclusively about Facebook, a front-page article in the November 15, 2018, edition of *The New York Times* is responsible for one of the most powerful exposés to date about the dangers posed by tech.[4] From Cyber Bullying, to fake news, the unauthorized selling of our personal information to nefarious third parties, Russian interference with the 2106 US elections, Facebook not only knew about these and other serious problems, but deliberately ignored clear and persistent Early Warning Signs that they were guaranteed to occur—indeed, were already occurring. (With few exceptions, a persistent trail of Early Warning Signs and Signals precedes all crises. If they can be picked up and acted upon in a timely manner, then many crises can be prevented before they occur, the best form of CM. In this way, Signal Detection is one of the key components of Proactive CM. We say more about this later.) To make matters worse, instead of taking quick and decisive action, it aggressively suppressed the signals. In essence, it didn't want to hear any bad news, let alone act on it. All of this was despite the fact that subordinates tried repeatedly to warn Facebook CEO Mark Zuckerberg and COO Sheryl Sandberg that major crises that could literally affect millions, if not billions, of users were virtually assured to occur. Far worse, Facebook engaged in deliberate smear campaigns to attempt to silence its critics. It shamefully portrayed prominent detractors as anti-Semitic. And, it lobbied major democratic and republican members of Congress to go easy on regulations. Once again, they are now so angry that have imposed major fines on Facebook and are pressing for tough regulations.

One of the Greatest Threats

To reiterate, technology, which has made our lives incomparably better in every conceivable way, now constitutes one of the biggest threats facing humankind. It threatens not only our physical, but our mental and social well-being. It forces us to confront some of the most important questions facing humankind. What are the *extent* and the *depth* to which we are willing to allow technology to intrude into our lives? In other words, how *pervasive* and *invasive* are we willing to let technology be? Further, because we *can* do something, does it mean that we *ought* to do it? Such questions are profoundly Ethical, for contrary to the outrageous contention that machines should be the new God, it is we who should control technology for our benefit and not be subservient to it, let alone elevate it to the status of a God.

With regard to the extent, i.e., the pervasiveness and thus the impact, of technology on our lives, an article in *The New Yorker* noted ominously:

---

[4]Frenkel et al. [5].

…One study has estimated that by 2030 the 'robocalypse" [i.e., the use of robots] will erase eight hundred million jobs.[5]

With regard to invasiveness, along with Elon Musk, AI enthusiasts have proposed putting chips directly in our brains so that we'll be "improved and enhanced humans," thereby better able to keep up with the increasing demands of modern life. As a result, AI will not only literally be *"in* us," but be an integral part *"of* us." The hope is that we'll thereby be more able to communicate seamlessly with all of our marvelous devices. Whether intended or not, the direct consequence is that we'll be cyborgs. The lines between humans and machine will have become permanently, if not irreversibly, crossed, if they have not already essentially vanished altogether. We're already so addicted to our innumerable devices such that for all purposes they are inseparable parts of us. In the not so distant future, the ultimate question will be, "Who and what will be human?" To put it mildly, it's an Ethical and Moral question of the highest and most pressing order.

Even without putting chips in us, there is mounting evidence that prolonged exposure to Social Media adversely affects the development of brains in young children. It also affects adults as well.

Nonetheless, we would be remiss if we didn't acknowledge that there is a persistent, ongoing debate as to how much Social Media are responsible, if at all, on the attitudes and states of mind of young people.[6] In part, it's due to the sharp differences in methods and underlying philosophical positions between those who favor large-scale surveys versus those who are partial to one-on-one clinical observations.[7] Thus, large-scale surveys do not necessarily find that Facebook is directly responsible for mental health issues such as depression. Rather, the effects are more insidious such as the disruption of sleep. Nonetheless, even surveys find that Social Media is an important factor in Cyber Bullying, and in this sense, it affects the mental health of children. In addition, by means of clinical observations, it's been found that Social Media *are* a major threat to normal child development.

As always, the supreme challenges facing us are not technical per se, but profoundly Ethical. They are embodied in the justifications we give for allowing a technology to proceed in the first place, let alone the Ethical ends to which it's intended to serve. To make no mistake about it, technology is never neutral. Even though it's often used for purposes and in ways other than those for which it's intended, it's always designed with some supposedly desirable end in mind. We say more about the Ethics of tech later.

In proposing that chips be placed in our brains for the prime purpose of *our* engineering the next stages of human evolution, little consideration has been given to what's to prevent someone from hacking into the chips and thereby gaining access to our most personal thoughts and desires, if not directing them altogether, at

---

[5]Friend [6].

[6]Denworth [7].

[7]See   https://www.npr.org/2019/08/27/754362629/the-scientific-debate-over-teens-screens-and-mental-health?fbclid=IwAR3pnRmJYECCONpHZWQ50uylBMVrDUllMK9boLfXFJtnBr0T jM4WEUyFfJs; see also Turkle and Together [8].

the very least seriously affecting them. How in other words do we protect ourselves against the unwanted intrusion of malevolent parties? If we have trouble protecting our data, how can we protect our minds and bodies?

The fact that the mind–body organism is a very complicated system that is highly dependent on the interactions between all of its parts such that no one can foresee, let alone predict, all the things that could go wrong don't seem to have crossed the minds of those making such proposals, or at least not to the extent that it should. Indeed, the mind–body organism is as complicated, if not more so, than any technical system currently in existence.

If this weren't ominous enough, efforts are under way to build robots that can not only read, but respond to our emotions. Apparently, more and more of us feel more comfortable in talking to an AI-enhanced robot about our deepest feelings and emotional states than we do to another fellow human being. Again, because we can do something, does that mean that we should necessarily do it? "How will it impact

Table 1 Threats posed by technology

Threats to democracy/truth/collective trust/society
  Mis- and disinformation: corrosion of trust/belief
  Platforms for fake news: corrosion of trust/belief/misuse by authoritarian regimes
  Interference in elections: integrity/fairness
  AI-alternate realities/videos: who and what is real/believable?
  Corrupt business models: unwarranted collection and selling of personal data to unscrupulous third parties
  Unregulated monopolistic/predatory power of big tech companies
  Collection of huge amounts of personal data by big tech companies
  Lack of business ethics/Crisis Management
  Lack of appropriate government regulations/fighting regulations
  Surveillance capitalism

Existential threats to the self/well-being/
  Repeated assaults on one's beliefs/values/sense of self-worth/personal and collective sanity
  Redesigning humans: altering genetic codes, AI chips in brains
  Directing evolution: Who and what is/will be human?
  Personal privacy/integrity: unwarranted collection and selling of personal data
  Loneliness/isolation/depression with increased use of Social Media
  Cyber Bullying
  Inability to initiate/sustain unscripted conversations
  Use of facial recognition technology to prey on young children
  Unwillingness of tech companies to plan for the unanticipated effects/abuses/misuses of technologies
  Lack of appropriate government regulations/fighting regulations

Remedies
  OTA/office of technology assessment: OPFT/office for the protection from technology
  Technology court: pros versus cons of a proposed technology
  Proactive Crisis Management
  Increased government oversight/regulations/EU model
  EU model: ombudsman for children
  Multidisciplinary teams of parents/social scientists/educators/kids/technologists

human beings and society in general?" needs to be given prime consideration. And, "How will it broadly affect our relationships with our fellow humans?"

Table 1 presents a summary of the various threats posed by technology with which we deal throughout. It also includes a brief synopsis of various remedies.

Pervasiveness and Invasiveness

While we explore more examples throughout, the preceding is sufficient to illustrate two of the major themes of the book. The first, i.e., Robocalypse, is an instance of the *pervasiveness* of technology, the extent to which it affects society as a whole. The second, AI chips in our brains, is an example of *invasiveness*, how deeply it intrudes into our minds and bodies, and thereby also affects us as a whole. Since the two constantly interact and therefore reinforce one another, they act on multiple levels of society simultaneously.

Pervasiveness and *invasiveness* are in fact two of the major dimensions that are critical in evaluating the threats posed by technology. For example, Facebook scores poorly on both counts. It's pervasive with regard to its effects on society as a whole: its ability to spread and to serve as a platform for fake news, dis- and misinformation, interference in our elections by nefarious foreign governments, etc. It's nothing less than a direct threat to our democracy (see Table 1). It's invasive in that once again, a growing body of studies shows that the more the young people use Facebook, the lonelier, more isolated, and depressed they are. If this weren't bad enough, Social Media have seriously impacted the ability of young people to engage in spontaneous, unscripted conversations.[8] For there are no "strict rules" as it were for starting, maintaining, and ending conversations.

In particular, the American Psychological Association has shown that adults suffer as well from anxiety and depression the more they use Social Media.[9] The supreme irony is that Social Media that was supposed "to better connect and bring us together" is one of the greatest threats to our ability to relate socially to one another.

Other dimensions play an equally important role in assessing the threats posed by technology, and thus hopefully, in our ability to control it for our benefit: (1) whether the potential dangers of a particular technology are *preventable* and (2) whether they are *reversible*. They are important with regard to whether we ought to go forward in the development and subsequent deployment of a technology. Stronger still, they *need* to play a major role. Thus, if we go forward and later find that a technology is harmful, are the effects reversible or not? Obviously, those things that pose grave danger but are neither preventable nor reversible are extremely serious and thus need to be strictly controlled.

---

[8]Turkle, *Ibid.*
[9]Rebecca A. Clay, *op cit.*

One of the most disturbing cases is the following. Although it was done for the ostensible purpose of eliminating the threat of childhood diseases, Chinese doctors have made direct alterations in the DNA of twin girls thus giving rise to genuine fears of "designer babies." It's a prime case of something that should have been debated seriously, if not prevented altogether, but was not. Even worse, the deeper fear is that it's not reversible. We discuss such considerations further.

We cannot emphasize enough that left unchecked technology constitutes an existential threat of the gravest order. Not only are society and democracy under continuous assault, but so is the fundamental nature of the self. We cannot overemphasize the dangers inherent in the fact that for better and for worse, we possess the godlike ability to intervene at the genetic level and thus alter the basic makeup and thereby the ultimate nature of humans.[10]

Beyond Privacy

Because of its importance and the constant attention that is rightly paid to it—indeed, that it demands—we need to say a few words about privacy. Privacy is in fact the single issue that's most responsible for the growing doubts about the benefits of technology and thus the rising backlash against it. It also illustrates many of the issues connected with pervasiveness, invasiveness, preventability, and reversibility.

To our detriment, the USA has among the weakest safeguards in the world when it comes to protecting the data that users both willingly and unwillingly give to tech companies every hour of every day. And, as we've noted, the USA also has the weakest protections in safeguarding young children from harmful content.

Once people check the "agree box" on an app or program, they essentially give tech companies the unfettered right to use their data with few if any restrictions. In addition, they have virtually no understanding of how and by whom their data will be used and for what purposes. The result is that there are little reasons to trust tech companies to protect us especially when their profits are tied directly to selling our most personal and highly sensitive information to third parties, with once again Facebook being the preeminent example.

In sharp contrast, the EU and the UK in particular have some of the strictest privacy protections of all. First of all, the agreement statements must be written in plain easy-to-understand language that as much as possible is devoid of unintelligible legal jargon. Second, users must be informed on a constant and timely basis of how their data is being used. In other words, they must give their explicit consent as to how, when, and by whom their data will be used.

The time has come to apply the same tough standards to the USA. Weak, after the fact, and self-regulations just don't work. We have to get out in front of potential breaches of our most personal data before they are too onerous to fix.

---

[10]Doudna and Stern [9].

The situation has gotten worse given the fact that tech companies now collect enormous data about us, as much if not more than the government. Accessing our credit information, social security numbers, where we live, work, etc., without our full knowledge and explicit consent—if not stealing them outright—are bad enough, but it now includes obtaining our preferences for products, what TV shows and movies we like and don't like, what books we read, for whom we vote, access to our deepest values and beliefs, and so forth. Yes, they have now agreed reluctantly to protect personal data such as credit information, but they have not taken steps to safeguard other data that they have skimmed without our awareness, let alone our full consent. As a result, traditional privacy agreements fall short.

For one, the users of apps need to be made more aware of how they can control the privacy settings on their devices. Information regarding their location, the phone calls they make, sites accessed, etc., can be controlled. As such, users can prevent third parties from gaining access to their personal information by controlling the privacy settings on their phones, thereby blocking unauthorized parties. There need to be clear instructions for "opting in" versus "opting out."

But even more, we need to be reassured that tech companies are doing everything they can to protect us from how *all* the data pertaining to us will not be misused or abused by either the company or third parties. Precisely because they have resisted such steps so strongly in the past is exactly why tech companies need to be strictly regulated. They have spent millions of dollars in lobbying to help ensure that they get the weakest regulations they want. For this reason, we propose that before they are allowed to operate, they must first develop and then submit plans to a new government agency, about which we talk later, charged with protecting the public from the dangers attendant to all technologies. In other words, tech companies must pass strict tests before they are given licenses to operate. They must be rebranded for what they are: "media companies" that are subjected to strict government regulations.

Writing in *Time*, Richard Stengel has proposed the creation of a senior federal official and even a cabinet office to deal with disinformation and election interference.[11]

If "scrubbing our data" isn't worrisome enough, consider once again that more portentous things are in the offing such as putting chips in our brains so that we'll be "improved and enhanced humans." We cannot overemphasize: What's to prevent someone from hacking into the chips thereby gaining access to our most personal thoughts and desires?

Figure 1 not only summarizes the discussion thus far, but illustrates the intense interactions between all of the various factors.

In sum, tech needs to be strictly regulated! It cannot be allowed to operate at will.

---

[11]Stengel [10].

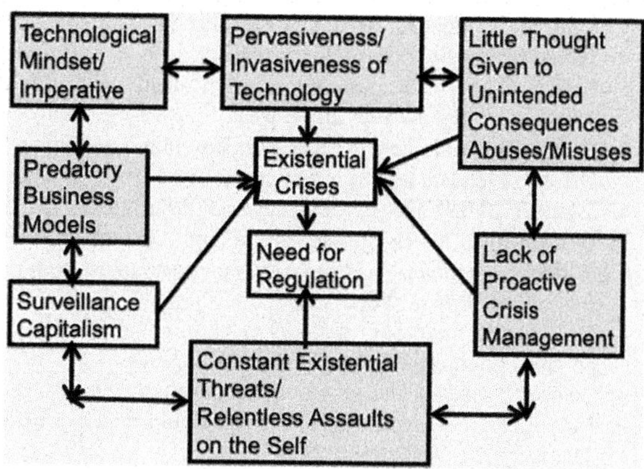

Fig. 1 Interactive effects of technology

In closing, a recent article in *The New York Times* shows that the fears regarding technology are far from being overblown.[12] By use of facial recognition technology, it's now possible to match and thus identify the faces of those who have had magnetic resonance imaging or MRI of their brains without their knowledge and thus consent. The inescapable conclusion is that for all purposes, Privacy Is Dead!

Concluding Remarks

This book is about what needs to be done to ensure that technology serves us, not for us to be subservient to it.

Because we have no way of knowing for certain whether the potential threats that we and others have identified—even more worrisome, those that we have not—will turn into full-blown crises and calamities is precisely why we are sounding repeated words of alarm. Before one can mitigate and thus hopefully prevent the threats from turning into full-blown, out-of-control crises, one first has to acknowledge their possibility and monitor their status carefully. In our experience, the numbers of organizations that acknowledge threats and prepare seriously for them are far too few. The time to start preparing is way overdue!

But even more, we have to acknowledge that many of the threats that we and others have already identified are taking place.

References

1. Broder-Sumerson J, Clay RA (2018) Treating the misuse of digital devices: psychologists are spearheading efforts to help wean people off technology. Monit Psychol Am Psychol Assoc 49(10)
2. Duhigg C (2018) Stop, thief. The New Yorker, 22 Oct 2018, p 58

---

[12]Rolta [11].

3. Singer N, Conger K (2019) Google fined $170 million for tracking of children. The New York Times, Thursday, 5 Sept 2019, p B1
4. Lohr S (2019) New inquiries show big tech scrutiny is rare bipartisan act. The New York Times, Saturday, 7 Sept 2019, p B1
5. Frenkel S, Confessore N, Kang C, Rosenberg M, Nicas J (2018) Delay, deny, deflect: how Facebook leaders leaned out in crisis. The New York Times, 15 Nov 2018, pp A1, A14–A15
6. Friend T (2018) The political scene: GOLDEN BOY 2.0. The New Yorker, 5 Nov 2018, p 22
7. Denworth L (2019) The kids are all right. Scientific American, Nov 2019, pp 45–49
8. Turkle S, Together A (2011) why we expect more from technology and less from each other. Basic Books, New York
9. Doudna JA, Stern SH (2018) A crack in creation: gene editing and theunthinkable power to control evolution. Mariner Books, New York
10. Stengel R (2019) The global war on truth. TIME, 7 Oct 2019, p 39
11. Rolta G (2019) Experts fear software could match you to your MRI. The New York Times, Thursday, 24 Oct 2019, p B8

> *Let's play a game I like to call 'This Gadget's Worst Nightmare.'*
> *The game goes like this: Pick some newish tech thing. Then spin the story out in your head. Imagine all the ways the technology might be misused, and ask yourself if our laws and politicians are prepared to handle the dystopian nightmare that emerges. I'm going to warm you: The game gets scary very quickly.*
>
> *Manjoo* [1]
>
> *As someone who studies the impact of misinformation on society, I often wish the young entrepreneurs of Silicon Valley who enabled communication at speed had been forced to run a 9/11 scenario with their technologies before they deployed them commercially.*
>
> *Wardle* [2]

With this chapter, we begin the development of a Theory of the Unthinkable. The starting point is the recognition that technology—more accurately, technologists—generally promises more than it can possibly deliver. Indeed, unbridled promises and unbounded enthusiasm are part and parcel of the basic makeup of the inventors of all goods and services. Needless to say, such promises wouldn't exist were it not for our insatiable need to satisfy our wildest, and endless, dreams and fantasies.

In return for the promise/desire for unrivaled beauty; instantaneous and constant connection with an endless array of friends; unsurpassed and lasting fame; immediate gratification of our every wish and desire; unparalleled popularity; superhuman power and strength; and indeed, defeating death itself,[1] we've sold our souls to the great God—or better yet, Devil—of technology. But it's a false bargain for

---

[1]See Harari [3].

© The Author(s) 2020

I. I. Mitroff and R. Storesund, *Techlash*, Management, Change, Strategy and Positive Leadership, https://doi.org/10.1007/978-3-030-43279-9_2

technology can never fully deliver, and hence satisfy, our starry-eyed dreams and hopes.

Fundamental to Thinking the Unthinkable is the consideration of as many ways as possible as to how satisfying our unbridled dreams and hopes by means of technology leads to not only severe disappointment and collective distrust, but worse yet, major crises. A major factor is the all-too-common failure to think about the actual contexts in which technologies are used versus the simplified and idealized ones that are typically presumed. We say more about this later in this chapter and the next one as well. This chapter is only the beginning of a Theory of the Unthinkable. We introduce additional key elements as we proceed.

To reiterate, it's part of the basic nature of technologists—if not the inventors of all goods or services—to see mainly, if not solely, the positive benefits of their marvelous creations and thus to downplay and ignore altogether the negative consequences and disbenefits. Facebook is a prime example par excellence. It was ripe for systematic abuse and misuse. Indeed, if we had deliberately set out to design a specific instrument that could be systemically abused and exploited, we couldn't have created a better, more powerful, and dangerous mechanism. No wonder why increasingly, Facebook, and in particular its founder, are portrayed in the worst of terms.[2]

Indeed, cartoons in *The Week* feature absolutely horrid portrayals of Mark Zuckerberg. In one, he's shown driving a car with the license plate, "Facebook soliciting kids," along with the caption, "Hey, Little Girl! Twenty Bucks If You'll Show Me All Your Private Stuff!"[3] In another, he's standing between racists, lying pols, Nazis, and Russians along with the caption, "It's Hard To Deny Friend Requests…"[4] From the standpoint of Crisis Management, such portrayals are literally the worst nightmare for any individual and organization. They are virtually impossible from which to recover. As such, they are prominent examples of the Unthinkable. Who among us can envision us and our organization being shown in the most disparaging light, let alone how to withstand it?

While not as portentous, consider a more bothersome, and growing problem, electrified scooters. Their primary attraction is not only are they fun to ride, but they offer an easy way to get around crowded urban centers. The problem is that users are not generally responsible when they are finished riding them. Indeed, they are largely irresponsible. They just dump them as they are on sidewalks, thereby creating a hazard for passing pedestrians and especially the elderly. As such, their users are a primary example of arrested development, i.e., young teenagers who are oblivious to their surroundings.

As an important aside, by being largely unconcerned with the potential abuses, misuses, and unintended consequences of their creations, arrested development may be more common than we'd like to admit among young, primarily male technologists. It's thereby one of the greatest barriers to the establishment of Socially

---

[2]Manjoo [4].
[3]See *The Week*, February 15, 2019, p. 20.
[4]*The Week*, December 13, 2019, p. 20.

Responsible Tech Companies. It's further aided and abetted by the tremendous amounts of monies at stake. And, it's also exacerbated by the tight tech communities, or "bubbles," in which they work and live that subtly and not so subtly encourage its members to behave and think alike.

To return to scooters, the failure to take into account the actual ways in which they would be used is responsible for the conspicuous threat they pose. No wonder why many cities have considered banning them outright.[5] The point is that technology always impacts, and in turn is impacted by, the larger social environment in which it operates. The failure is not talking into account the broader contexts in which all technologies are used and impact.

If technologists and tech companies are largely insensitive to the unintended social consequences of their prized creations, how then does one foresee and take appropriate steps to protect us from the dangers that not only are lurking in, but result from all of our creations? By continuously and systematically Thinking the Unthinkable! As we discuss later, it needs to be a fundamental part of a new federal office of socially responsible technology or OSRT. It also needs to be a basic part of the job of the chief technology officer who is responsible for the use and development of technology whether the organization is directly involved with the production of technology or not. Indeed, it's hard to imagine any organization that is not heavily involved with technology in one way or another. And, the chief officers for Crisis Management and Strategy need to be directly involved as part of a close-knit corporate team.

As we indicated in the Preface, Denmark has taken the groundbreaking and unprecedented step of appointing an ambassador to Silicon Valley to represent its interests against the giant tech companies.[6] It did this after it determined that "tech behemoths now have as much power as many governments—if not more."

Thinking the Unthinkable requires us to attempt what few are inclined to do, let alone are capable: to imagine how every one of the proposed positive attributes and benefits of a technology can lead and/or turn into their exact opposite. Such a task is virtually impossible for single individuals to accomplish entirely on their own. Thus, shortly after Facebook's release—ideally prior to it—it would have required teams of parents, psychologists, teachers, and even kids themselves to envision how it could serve as a major platform for relentless Cyber Bullying. Indeed, in retrospect, such a scenario was virtually guaranteed to occur and, in this sense, was perfectly predictable, that is, if one had wanted to think systematically and comprehensively about such things in the first place.

To reiterate, to foresee, and to guard against the dangers lurking within a technology require us to do one of the most difficult things with which humans are charged: thinking of as many ways as possible to undermine our most cherished creations.

One of the most powerful ways of Thinking the Unthinkable is by uncovering the host of taken-for-granted assumptions that are made about the larger body of

---

[5]Griffith [5].
[6]Stariano, op cit.

stakeholders who use and are affected by a technology, and then challenging those assumptions as ruthlessly as we can. For example, a commonly made assumption about users is that they are conscientious and their intentions are honorable. In short, they want to do the right thing. As a result, they will follow scrupulously the instructions for how to use a technology responsibly. Accordingly, they will not abuse and misuse it for malicious ends like Cyber Bullying or interference in our elections.

Thinking the Unthinkable also requires something even more difficult: the emotional fortitude to face up to tough problems long before they occur. For this reason, one cannot emphasize enough the front-page story in the November 15, 2018, edition of *The New York Times* that showed how Facebook not only ignored, but deliberately denied clear and persistent warning signals of major problems.

A Theory of the Unthinkable: Part I

The following are some of the key elements of a Theory of the Unthinkable. We introduce additional elements as we proceed:

1. Major unanticipated and/or systematically ignored stakeholders who can and will deliberately cause harm, at the very least interfere with the ideal and idealized aims of a technology;
2. Major idealized attributes/assumptions about principal stakeholders that are later proven to be false;
3. The idealized contexts in which a technology will be used and which later prove to be false as well;
4. Idealized assumptions about the key benefits/properties of a technology and how the exact opposite can and will occur;
5. Idealized assumptions about how one can contain the damage, if any, if a technology fails, and worse, is responsible for widespread harm; how such assumptions are demonstrably false;
6. Idealized assumptions as to why there will not be widespread calls for the strict regulation, and worse, the strict banning and/or elimination of a technology.

We say a word about each.

First, "stakeholders" are all the parties who affect and are affected by the invention, licensing, marketing, operation, regulation, etc., of a technology. In short, they are deeply involved with every aspect and phase of a technology from its initial conception and development, operation, to its eventual obsolescence and replacement. For instance, a basic taken-for-granted assumption is that malicious, nefarious actors will not abuse and/or misuse a technology for their ill ends alone. In other words, the major players/stakeholders are not only known, but benign and responsible.

It's also generally assumed that the contexts in which a technology will be used are known, and predictable, benign, stable, and under control. For instance, Social Media will not be a major factor in the spread of disinformation, misinformation, hate speech, etc.

For another, technology will work as promised and its benefits are clear and demonstrable. It will certainly not be responsible for major harm or injuries. It will work as planned because its properties are essentially known and predictable. In other words, there will be no major surprises.

Since it's generally assumed that a technology will work as intended, and its benefits are clear and incontestable, there is essentially no need for backup plans for CM since failure is essentially precluded. In essence, planning for the worst is nothing but a waste of precious time and resources. In the same way, there will not be major calls for its elimination or strict control.

Stakeholders

While we discuss each of the preceding elements in more detail throughout, we want to close this chapter with a more thorough examination of the first factor. In particular, we want to show how one can think systematically about the stakeholders and one typically doesn't take into account and indeed, often, doesn't want to consider.

One of the best ways to think about the general kinds of stakeholders that are involved in any and all technologies is by means of expanding circles. In the innermost circle are all those parties who are involved in the creation, direct financing, leadership, and marketing of a technology, product, or service. It also includes all the immediate employees and supervisory personnel in the operations of a company.

In essence, assumptions are the presumed properties of stakeholders, what they are like and how they will behave. Thus, it's commonly assumed, if not taken for granted, that one's primary innermost stakeholders are capable, dedicated, dependable, honest, intelligent, loyal, trustworthy, etc. In other words, not only are they not malicious, but they are a demonstrable and valuable asset. They certainly will not cause or be responsible for harm in any way. Sadly, this is often not the case with disgruntled or disaffected workers who have engaged in workplace sabotage and violence. The trick is to be on the constant lookout for malicious internal actors without creating an oppressive/Orwellian-type workplace or culture that actually encourages precisely that which one wishes to prevent. One of the best, but not perfect, ways of doing this is to involve employees in how they will "police themselves."

In a later chapter, we explicitly examine the notion of harm. If they even consider it at all, most companies take the avoidance of harm for granted. Instead of considering it explicitly, especially from the standpoint of Ethics, it's "under the radar screen."

The next, outermost circle is composed of all those parties immediately external to an institution or organization, e.g., competitors, the media, regulators, and most of all, the direct users of a technology, product, etc. Once again, it's commonly assumed that users are competent, intelligent, etc., so that they will use a technology responsibly as intended. They will certainly not abuse or misuse it. Worst of all are external malicious actors who are deliberately out to cause as much harm as possible through the improper/unintended use of a technology and whom rarely are considered explicitly and systematically.

This latter circle also includes the most vulnerable users and those who are most affected by a technology. Namely, what is benign to one party may be harmful to others. Similarly, not all are equally affected by what are deemed unintended consequences and side effects. In other words, what's a positive benefit to one class of stakeholders is not necessary to others. We talk more later about a specific process for addressing this important issue.

Outer circles contain foreign actors, governments, etc.

Many more kinds of stakeholders exist that are not typically included when planning for a product or service. The following is a specific example of how they not only need to be addressed, but can be.

A company that produced a product that was intended only for adults discovered that young children were using it in ways that raised concerns. Somehow or another, the company realized that if young kids were using their product, then they needed to hire an expert in child development to give them continuous ongoing advice of how they could redesign the product in ways so that it would not pose an immediate danger to kids. Given that all technologies now affect children and in essence are used by everyone, all tech companies need to expand the base of their "normal personnel."

In addition to considering explicitly as many parties as possible that are involved in the invention, promotion, distribution, etc., of a technology or product, one especially needs to consider those who will use it to further nefarious ends. Stakeholders and the associated assumptions we are making about them are among the most important factors in Thinking the Unthinkable.

A comprehensive list of stakeholders while extremely important is only one factor in Thinking the Unthinkable. We not only need to say more about all of the other elements, but even more need to demonstrate a process that will help ensure that we have given them the serious consideration they require and, thus, that they will be acted upon in the right ways to ensure the health and well-being of all those who are affected in one way or another by technology.

References

1. Manjoo F (2019) Yes, that doorbell is spying on you and it's time to panic. The New York Times Opinion, Sunday, 14 Apr 2019, p G6
2. Wardle C (2019) The new world disorder. Scientific American, Sept 2019, p 88
3. Harari YN (2017) Homo deus: a brief history of tomorrow. Harper, New York, pp 22–23

4. Manjoo F (2019) Zuckerberg should just retire. The New York Times, Thursday, 24 Oct 2019, p A23
5. Griffith E (2019) San Diego's scotter tryout gets off to a bumpy start. The New York Times, Wednesday, 4 Sept 2019, p B1

Pluses and Minuses

> *…every time [Mr. Schroepfer, Facebook's person in charge of erasing millions of unauthorized posts] and his more than 150 engineering specialists create A.I. solutions that flag and squelch noxious material, new and dubious posts that the A.I. systems have never seen before pop up—and thus are not caught. The task is made more difficult because 'bad activity' is often in the eye of the holder and humans, let alone machines, cannot agree on what that is.*
>
> *Metz and Issac* [1]

The ability to foresee as many of the potential dangers as possible lurking within a technology is one of the key factors in Thinking the Unthinkable. As we've stressed, one of the best ways of accomplishing this is by listing as many of the supposed benefits of a technology as one can and then considering as many ways as possible how the exact opposite can and will occur. In other words, what are all the ways in which technology proposes to make our lives better, and then, how can it systematically fail to do so?

The starting point is the long-standing, general aim of technology: "Technology not only magnifies the senses, but it allows us to surmount the limitations of our minds and bodies." Technology thereby allows us to hear, see, and sense things we would not be able to do without it. It allows us to communicate with others over long distances. It permits us to be in touch instantly with unsurpassed numbers of others at a moment's notice. It lets us travel long distances comfortably and safely at speeds once thought to be beyond the bounds of human possibility. It now promises to extend the power of human thought indefinitely. More portentously, it promises to extend human life endlessly, if not defeat death altogether. It can sense when our bodies are in danger of being harmed and then offer needed protection. And, it promises to capture and to respond to our deepest, most personal inner feelings and emotions.

© The Author(s) 2020
I. I. Mitroff and R. Storesund, *Techlash*, Management, Change,
Strategy and Positive Leadership, https://doi.org/10.1007/978-3-030-43279-9_3

But if technology brings great gifts and grants our miraculous wishes, it also poses equally portentous threats. If robots promise to do much of the work we find onerous and dangerous, they also threaten to relegate humans to permanent subservience and irrelevancy. Artificial Intelligence or AI offers not only to augment our mental and emotional capabilities, but also to replace us altogether. While driverless cars may be safer in the long run, how do we cope with the millions who will lose their jobs and dignity as a result of not being employed?

Consider that the heightened ability to hear and see objects and events that we could not without technology also makes possible the increased ability to spy on our most personal and private conversations and to relay them to others without our explicit knowledge and consent; witness Alexa and Echo.

Again a growing body of studies is testimony to the fact that instead of contributing to and thus raising the self-esteem of young people and adults, the more the people use Social Media, the more it lowers their self-esteem. After all, they are constantly comparing themselves against all of the idealized portraits of others that they cannot ever hope to match. It's a losing proposition that produces noticeable psychological damage.

Social Media are the preeminent example of technology producing the exact opposite of what was intended. The very thing that was supposed to bring us closer together is now one of the biggest factors responsible for driving us apart.

To reiterate, key to Thinking the Unthinkable is taking every one of the proposed benefits of a technology and then showing how and why their exact opposite can occur. Given its extreme importance, we discuss in a later chapter a detailed process for accomplishing this. It involves among other things the simultaneous consideration of all of the factors that are essential in Thinking the Unthinkable. That is, it's not possible to separate a discussion of the idealized properties and supposed benefits of a technology from the idealized contexts in which it is supposedly used. Any discussion of the positive benefits of a technology implicitly presupposes any number of idealized contexts or settings.

However, before we can discuss a process that allows us to consider all of the factors simultaneously, we need to discuss a special way of Systems Thinking, which is the topic of the next chapter.

But first, we want to discuss the limitations of AI. It's one of the most important examples of why technology not only fails to achieve its desired aims, but results in the exact opposite.

The Fatal Flaws of AI

AI rests on a central premise: that all of human behavior—thought itself—can be not only be captured, but reduced to sets of rules—the fancy name is algorithms. By putting algorithms into computers, the prime contention is that they will not only speed up, but lead to better decisions. It doesn't matter if the "rules" are based on probabilities of what we and others are mostly likely to do. It doesn't even matter how many of them there are. All that counts is that they can be captured and encoded into algorithms.

The preceding is not only fundamentally misleading, but dangerously false.

AI rests on other key premises as well. For instance, by examining hundreds, if not thousands, of cases and thereby feeding huge amounts of data into so-called deep learning machines, they will be able to discern how humans learn. First of all, this ignores the fact that even young babies are able to learn from just a few, messy cases, not hundreds of them.[1] The key question is how humans, both rightly and wrongly, learn from a few cases, not many of them. The short answer is that they rely on "approximate rules of thumb" or heuristics that do not guarantee success but "work well enough."

Second, it also ignores that thoughts—the act of thinking itself—do not exist by themselves but are parts of a complex mind–body system. In short, the body is not just a "fancy or sophisticated carrying case" for the brain. Instead, the mind is distributed throughout the entire mind–body system. There is in fact a kind of "primitive brain" that surrounds the heart and is essential to its functioning. Capturing thoughts thus entails capturing, if only in part, the states of the entire mind–body system, at the very least, many of the most important interactions between them. In other words, the "contexts" in which tech is used and affects us are much broader than those that are typically assumed. There are no such things as "standalones," i.e., completely self-contained contexts and/or situations.

It also ignores the fact that the notion of "mind" is fundamentally "social." There is no such thing as a completely isolated, self-contained, and individual mind.

Third, the idea that thoughts can be reduced to rules alone ignores the basic fact that that thoughts and emotional states are inseparable. There are no thoughts without accompanying emotions and vice versa. And, emotions are not subject to the same kinds of rules. Indeed, much of our emotions and thoughts are not available to consciousness. They are triggered by other emotions, events, and thoughts of which we are only dimly aware, or at the very least, not fully. They are influenced and part of our basic hopes, dreams, fears, and anxieties.

In addition, Emotional Intelligence—knowing how to pick up and respond appropriately to the emotional cues and states of others as well as oneself—is very different from Cognitive Intelligence.

Fourth, and extremely important in its own right, the whole notion of algorithms that are collections of clear-cut, consistent rules ignores the basic fact that there is no aspect of human behavior that is not subject to differing if not contradictory views and opinions. Medicine is a prime example. True, more and more of medicine is evidence based. But this doesn't mean that doctors evaluate the "same situation" exactly alike. After all, they bring different experiences, judgments, and intuitions to bear on everything they do.

---

[1]See Gopnik [2].

Gary Marcus and Ernest Davis put it as follows:

> …It's the set of background assumptions, the conceptual framework, that makes possible all our thinking about the world." Therefore, in order for AI to work, it's essential to capture the background assumptions that are a crucial part of everything we do. Nonetheless, as they state, "Yet, few people working in A.I. are even trying to build …background assumptions into their machines…we're not to get sophisticated comport intelligence without it."[2]

Consider one of the most contentious and problematic situations: "determining who is and is not a 'good, responsible' potential employee." Such judgments are historically full of enumerable ethnic, racist, and sexist biases so that basing algorithms on "normal standard practices" is not only misleading, but highly Unethical.[3]

Hiring the "right person" is not just a matter of selecting someone with the "right credentials" such as whether he or she went to the "right school," majored in the "right subjects," got the "right grades," etc. It's as much the case whether one is a "good fit" with the groups with which he or she will be working. And, it invariably involves whether a company needs to undergo serious Diversity Training so that it confronts its underlying biases and thereby expands its taken-for-granted notions regarding whom it needs to hire, let alone promote.[4]

An equally, if not more, critical case is that of health care. If one uses historic data on which to build algorithms to determine who should receive the proper amounts of health care, then Black people typically fare poorly and as a result become sicker. Historically, substantially less money has been spent on the health of Blacks than for Whites. Thus, using the cost spent on care in the past is a poor proxy for health needs. It shows that using the "right category or label" is critical for "solving the 'right problem.'" It also shows that using "the right label" is a matter of "critical judgment" that only humans can exhibit, not machines.[5]

All data are subject to different interpretations. It's rarely the case that one position is "completely right" and the other is "completely wrong." Complex decisions are not like simple, canned exercises for which there are only "single right answers." We say more about this in the next chapter.

F. Scott Fitzgerald said it best: "The test of a first-rate intelligence is the ability to hold two opposed ideas in mind at the same time and still retain the ability to function." To put it mildly, this is a very different notion of intelligence from that which underlies current thinking about AI.

Humans are not bundles of consistency through and through, but of the constant ebb and flow—enduring struggle—between opposing thoughts and emotions. Thus, if AI is truly to advance, it will have to take a radically different course. It will not only have to embrace, but deeply incorporate Dialectical Thinking, i.e., the strongest arguments pro and con that can be made for a particular position. In other

---

[2]Marcus and Davis [3].
[3]Ajuwa [4].
[4]Metz [5].
[5]Benjamin [6].

words, Dialectical AI is absolutely essential if we are truly to capture the intricacies and subtleties of human behavior and thought. We're waiting for evidence that the community of AI is ready in the slightest to do so.

As we show later, the ability to embrace Dialectical Thinking is one of the key attributes of the Socially Responsible Tech Company.

The following story while most likely apocryphal is one of the most powerful examples of the need for Dialectical Thinking. It shows in no uncertain terms that doubt is not only the underlying basis of Dialectical Thinking, but thereby one of the most critical attributes humans possess.

In the 1950s, at the height of the Cold War between the USA and the USSR, an Airman stationed in Alaska was seated in front of a large computer-like console. His job was to monitor it for swarms of incoming objects that showed in no uncertain terms that the USSR was about to attack the USA by means of Intercontinental Nuclear Ballistic Missiles. And sure enough, as if on cue, the console sounded an alarm and was prepared to launch a counterattack but if and only if the soldier gave his final approval. Fortunately, for some reason, the Airman in charge didn't believe that the attack was real and thus didn't press the big red button that would have effectively started World War III and ultimately would have led to the utter annihilation of both countries, if not the entire planet. The "large object coming over the horizon" for which the system was not programmed to recognize was the Moon!

The fate of humanity literally rested on the doubts—the nagging gut feelings—and good judgment of one human being. Why would we ever trust any machine to do the same? We always need intelligent humans to oversee our supposedly "brainy learning machines." Show us the autonomous AI system that even comes close to incorporating "doubt," let alone "gut feelings."

On the other hand, why should something so critical as the launching of war be left to the judgement of one person? It surely required the consideration of a team that was schooled in the perils of Group Think, and thereby how to avoid it.

Finally, there is the matter of the broader harm done to democracy as a whole. As Stuart Russell, a computer scientist at UC Berkeley, puts it:

> [There's] a really simple example, where [an] algorithm isn't particularly intelligent but has an impact on a global scale. The problem is that the objective of maximizing [getting people to click on an app] is the wrong objective because the company—Facebook or whatever—is not paying for the externality of destroying democracy. They think they're maximizing profit, and nothing else matters. Meanwhile, from the point of the rest of us, our democracy is ruined and our society falls apart, which is not what we want...AI systems that are maximizing in a very single-minded way [a] single objective, end up having these effects that are extremely harmful.[6]

---

[6]Worthington [7].

Closing Remarks

One of the biggest factors responsible for the Unthinkable is the failure to consider that all technologies can be used—abused and misused—in ways that were not intended, let alone not considered at all, by their developers. Latent defects, certainly fatal flaws, can and will be exploited. As result, even the benign properties of a technology can be taken advantage for nefarious purposes. For this reason, we talk later about a deliberate process that is the best of which we know for thinking about how technology can be exploited.

References

1. Metz C, Issac M (2019) It's never going to go to zero. The New York Times, Sunday, 19 May 2019, p Bu 6
2. Gopnik A, https://www.wsj.com/articles/the-ultimate-learning-machines-11570806023
3. Marcus G, Davis E (2019) Build A. I. We can trust. The New York Times, Saturday, 7 Sept 2019, p A23
4. Ajuwa I (2019) Beware of automate hiring. The New York Times, Wednesday, 9 Oct 2019, p A29
5. Metz C (2019) AI learns lots from us. Our biases too. The San Francisco Chronicle, Saturday, 16 Nov 2019, p D1–D2
6. Benjamin R (2019) Assessing risk, automating racism. Science 336:421–422
7. Worthington L (2019) A race against the machine: Stuart Russell wants to harness AI before it's too late. Californian, Fall 2019, p 36

Context: Technological Messes

4

Citing fears that doctored videos of political candidates could be used to manipulated in 2020, a California lawmaker has proposed legislation to ban the release of so-called deepfake [sic] images before an election.

Gardiner [1]

The contexts in which technology exists and functions are vastly more complex than what's typically assumed. The problems of society do not consist of bounded, well-structured exercises. All technologies operate in complex settings. As a consequence, the problems of society are not like the majority of simple exercises that constitute the backbone far too many classes in the Engineering and Physical Sciences, not to mention K-12 education in general.

The following is unfortunately all-too-typical of the kinds of exercises that are the basis of the body of education: "$X + 6 = 11$; find X." It's certainly characteristic of the kinds of so-called problems with which earlier generations were presented. As such, it's not a problem in the true sense of the term. First of all, it's completely well defined—indeed, overly so—such that everyone is not only expected to accept as one of the key building blocks of formal education, but secondly, applying the accepted rules of arithmetic, to get the single right answer "$X = 5$." In other words, exercises have one and only one "right answer" that everyone is expected to get. And, the procedures—algorithms—for getting the single right answers are clear-cut and unambiguous. Needless to say, a steady diet of exercises over a span of over 20 or more years makes many students extremely anxious and depressed when they have to confront real problems for which there are no simple, single, clear-cut answers. We know this from teaching many students and conducting numerous seminars over the course of our careers.

Real problems have none of the supposedly desirable characteristics of simple exercises. First of all, each stakeholder—once again, all the parties who affect and are affected by the problem—potentially sees it in very different ways. Thus, if

© The Author(s) 2020
I. I. Mitroff and R. Storesund, *Techlash*, Management, Change,
Strategy and Positive Leadership, https://doi.org/10.1007/978-3-030-43279-9_4

Sandra has an income of only $2000 a month, how does she feed and care for herself and her two children if her rent and other living costs come to $2500? The "answer" is not the simple arithmetic difference between $2500 and $2000. It's rooted far more in how one survives a precarious situation than it is an exercise in arithmetic. We wouldn't expect a social worker to see the problem in the same way that a parent, relative, spouse, or dependent child would. The point is that problem formulation and negotiation are two of the most important and crucial parts of problem solving. Before we can "solve" a problem, we first have to agree on its "definition."And in doing so, we have to ensure ourselves that we are not committing what's referred to as Type III Errors, "solving the wrong problems precisely." To minimize such errors, we have to engage multiple views of a problem and debate which ones best fit the situation. The earlier example of using cost as a proxy for health care is a prime case of "solving the wrong problem."

As a general rule, problems are parts of complex messy systems where the "basic problem" is not only to define the "fundamental nature of the problem," but its "relationship to the host of other problems to which it's connected in a myriad of ways." And, unlike simple exercises, Ethics plays a major role. Namely, why _ought_ we attempt to solve this and only this set of problems versus others? In the case of technology, who will benefit from it? Conversely, who will be potentially harmed? And, how can we anticipate such harm and what can we do to mitigate it?

Wicked Messes

The late, distinguished social systems analyst and thinker extraordinaire Russell L. Ackoff appropriated the word "Mess" to stand for a whole system of problems that were so interconnected—indeed inseparable—such that one couldn't take any single problem out of the Mess and attempt to analyze it, let solve it, independently of all the other problems without distorting irreparably both the basic nature of the so-called individual problem and the entire Mess of which it was a part. The notion of "individual problems" is nothing more than a highly misleading and outdated figure of speech. The interconnections between problems are as critical, if not more so, than the so-called individual problems themselves. For example, one can't even begin to formulate, let alone "solve," the problem of "homelessness" independently of "urban crime," "income inequality," "mental illness," "drug addiction," "civic indifference and opposition," and the host of other equally critical known and unknown problems to which it's connected in a myriad of ways, both familiar and unfamiliar.

A major consequence is that all of the various elements that constitute a Theory of the Unthinkable are parts of The Tech Mess. They neither exist nor function solely by themselves.

Messes are complicated even further by an additional confounding factor; "Wickedness." A "Wicked Problem" is a problem that cannot even be formulated by any of the traditional disciplines or professions acting singly or in concert with one another. In other words, Wicked Problems are "beyond" the scope of any of the currently known disciplines or professions. They defy—indeed, strongly resist—attempts to pin them down. In addition, a "solution," if it exists for one time and place is not necessarily one for all times and places. More often than not, so-called solutions are responsible for creating even worse problems.

Virtually all of the problems of modern societies are Wicked Messes. They are the epitome of unbounded, unstructured problems. Unlike exercises, Wicked Messes do not have simple, clear-cut, and stable solutions. One only "copes" with them as best one can. Yet, because of the historic influence of exercises, we persist in trying to pin them down and hence treat them as if they were exercises that had nice, neat, simple solutions.

In sharp contrast, Wicked Messes are only amendable to "approximate rules of thumb" or heuristics that allow us "to cope and manage them as best we can." For instance, a prime heuristic is "always be on the lookout for unanticipated—i.e., the most improbable–interactions among key factors." Another is "pay special attention to anything that threatens the most vulnerable stakeholders".[1]

The Tech Mess not only contains all of the elements that comprise a Theory of the Unthinkable, but all of the parties that are involved with any and all technologies. In this way, the various stakeholders who are charged with trying to cope with a mess are fundamental parts of it. So are all of the previous attempts in trying to cope with it. Thus, the associated history of a mess is a basic part of it as well. In fact, anything that is related to a mess is part of it. In short, Wicked Messes are the epitome of unbounded, unstructured problems.

One of the saddest and most disappointing aspects of the whole issue is that the acknowledgment of Wicked Messes is still rare. As a result, we are seriously lacking in research into the heuristics that allow us to cope with them. We can't think of a more pressing topic for research.

Concluding Remarks

AI and Wicked Messes demonstrate the limits of traditional Philosophic thinking. Conventional accounts of Empiricism and Rationalism are unable to cope with the demands of Wicked Messes. They demand a different way of thinking. We say more about this later.

Once again, Thinking the Unthinkable, and thus, coping with Wicked Messes, is a key part of the job of the senior most officer involved with ensuring the Social Responsibility of an organization.

---

[1]See Mitroff et al. [2], for an expanded list of heuristics for wicked messes.

References

1. Gardiner D (2019) 'Deepfake' videos would be outlawed by new bill. The San Francisco Chronicle, p D1

2. Mitroff II, Hill LB, Alpaslan CM (2013) Rethinking the education mess; A system's approach to education reform. Palgrave Macmillan, New York

> *Advances in computer vision have given machines the ability to distinguish and track faces, to make guesses about people's behavior and intentions, and to comprehend and navigate threats in physical environments.*
>
> *...I worry that we're stumbling dumbly into a surveillance state. And I think the only reasonable thing to do about smart cameras is to put a stop to them.*
>
> *Manjo* [1]

All of the rules and regulations per se will not protect us from the false promises and excesses of technology, let alone their negative and unintended consequences. What's needed is a carefully orchestrated, ongoing process that will continuously examine all of the elements, and especially their interactions, that are not only key to Thinking the Unthinkable, but even more, to coping with it. This chapter provides the broad outlines of such a process.

From 1972 to 1995, the Office of Technology Assessment or OTA was an agency of the US Congress. Its main purpose was to provide Congressional members and committees with supposedly "objective and authoritative analysis of complex scientific and technical issues." Over the course of its existence, it produced over 750 studies on a wide range of topics including acid rain, health care, and global climate change.

When OTA was closed in 1995, Republican Representative Amo Houghton criticized his own party for defunding the agency, noting that "we are cutting off one of the most important arms of Congress when we cut off unbiased knowledge about science and technology." Critics saw it as a prime example of politics overriding science, and numerous scientists called for its reinstatement.

While it was shut down in the USA, OTA survived, largely in Europe. Further, while campaigning for President in 2008, Hillary Clinton pledged to work to restore it if she were elected.

I. I. Mitroff and R. Storesund, *Techlash*, Management, Change, Strategy and Positive Leadership, https://doi.org/10.1007/978-3-030-43279-9_5

Today, OTA is needed more than ever. It not only needs to be resurrected, but rebranded as The Office for Socially Responsible Tech or OSRT.

While the Congressional Research Service has proposed re-establishing something like OTA, it does not go far enough. More is needed than an agency that merely gives "advice" to members of Congress on matters dealing with science and technology.

In particular, OSRT would especially subject those technologies which threaten to make significant alterations in the genetic makeup—thus at the level of DNA–of humans to serious inspection and control. They would not be allowed to go forward without its direct approval and continuous monitoring. Furthermore, given that technology is more encompassing than ever, unlike the old OTA, it would have no formal date of expiration.

Central to OSRT is the idea of a Science Court, or more appropriately, a Tech Court. First proposed in 1976, the Science Court never took off for a variety of reasons, mainly political. In effect, scientists on opposing sides of an issue would make their strongest case before a panel of specially trained Scientist Judges. Similar to a court of law, advocates would have the opportunity to question the evidence submitted by the opposing side. Having heard the evidence, the judges would render their decision. Further, it would be published so that the public at large would hopefully have a clearer understanding of the scientific issues at hand.

The Tech Court would work by having one or more sides argue the proposed benefits of a proposed technology, while others would not only argue the possible disbenefits, but how it could be systematically abused and misused as in the case of Social Media, certainly facial recognition technology. In short, the Tech Court is the living embodiment of Dialectical Thinking.

The Tech Court would especially scrutinize each of the elements of the Unthinkable to help ensure that they would not lead to major crises. It would pay especially close attention to the pervasiveness and intrusiveness of a technology, as well as to the preventability and reversibility of unintended consequences and undesirable side effects.

Most of all, OSRT would be guided by the Precautionary Principle. The burden would be squarely on the developers of a new technology to make the case that it would not cause significant harm to individuals and society at large. If they fail to do so, a particular innovation would be strictly prohibited, curtailed, and/or modified, assuming that it could be. If not, it would be withdrawn and eliminated altogether.

Needless to say, the Precautionary Principle has been subject to widespread criticism, the most prevalent being that it's biased against anything new because it imposes a severe threshold on anything truly innovative.

Be this at it may, when it comes to intervening in humans for the express purpose of "redesigning and thus improving the human condition," the Precautionary Principle must be given the highest priority. We should all be extremely wary of any and all proposals to "improve the human race."

In sum, more than ever, we need a new agency to protect us from the excesses of technology, and especially from the overinflated claims of their proponents. In order for regulations to do their job, they must be part of an ongoing process—a mechanism such as the Tech Court—that will ensure that they are appropriate and that a company is carrying them out as needed.

The Toulmin Argumentation Framework, TAF

The eminent Historian and Philosopher of Science, Stephen Toulmin developed an ingenious framework for analyzing the structure of arguments.[1] It's especially pertinent to the idea of the Tech Court. It's vital to Thinking the Unthinkable.

The Toulmin Argumentation Framework or TAF for short is deceptively simple. It consists of a Claim C, Evidence E, Warrant W, Backing B, and a Rebuttal R.

All arguments terminate in a Claim C or a set of Claims. The Claim is the end result or conclusion of an argument. With regard to tech, a typical set of Claims is one or more of the following: "Not only is our technology clearly superior, but it's a major step forward. It's unquestionably innovative. Indeed, it's revolutionary. Further, its benefits are not only crystal clear, but indisputable."

All arguments make use of some kind of Evidence E to support their Claim(s). Typically, E is whatever data or facts that one can muster that lend support to the Claim(s). In the case of tech, the Evidence is generally the Track Record of the Developers, i.e., their successes with previous innovations, how it's performed, etc. If this is their first venture, then the Evidence is typically their credentials, the testimony and support of peers, backers, teachers, etc.

The strongest E are facts from independent authorities supporting all of the proposed benefits and properties of a proposed technology. As with Claims, Evidence differs with regard to how strong or weak they are. Whatever the case, this is typically the Empirical Base of an argument.

In general, the different sides of an argument start with different Evidence because they are working backward from different Claims. A Claim can thus either be the beginning and/or end of an argument. Once one has a preferred Claim, one typically searches for Evidence to support it. As a matter of fact, in the vast majority of cases, one is generally working backward from a preferred Claim or set of Claims. The term that best captures this process is Confirmation Bias.

Those who are on the opposing side of a technology generally argue that "The benefits of a proposed technology are far from clear; indeed, not only does it lead to their exact opposite, i.e., dis-benefits, but to demonstrable harm." In short, "The reasons for going forward with the proposed technology are not acceptable, i.e., unsupported."

The upshot is that instead of facts or more generally Evidence always leading unequivocally to or driving conclusions, more often than not, it's the other way around. One starts with a favored, pet conclusion, or typically a set of conclusions,

---

[1]Toulmin [2].

and then works backward to make it appear that one derived it by first starting with "Impartial Evidence, Facts," etc.

The Warrant is the set of reasons why the Claim follows from the Evidence. By itself, Evidence does not directly imply a Claim. A good way to think of it is that the Warrant is a "conceptual or intellectual bridge" that allows one to go from a limited set of evidence E, to a general conclusion, C. To put it another way, the Warrant is the "because" part of an argument. For instance, a typical Warrant is: "Whenever E has resulted in the past, C has occurred _because_ E is not only an indicator of C, but a prime factor in its occurrence or causation; since E has occurred this time as well, we are Warranted in concluding C once again." In this particular example, the Warrant functions as a "continuity preserver." Supposedly, whenever a particular set of facts or certain events E, etc. have occurred "n" times, then we are Warranted in concluding that they will occur "$n + 1$" times. Furthermore, according to this line of thinking (argument), the larger the n the more we are entitled to conclude that $n + 1$ will result. Thus, if E has occurred 1000 times, then we feel confident that it will occur the 1001st time.

The problem of course is the fact that a rooster has been fed 300 days in a row is no guarantee that its head will not be chopped off on the 301st day!

If the Evidence is the Empirical Basis of an argument, then from the standpoint of Philosophic School of Rationalism, the Warrant is the Analytic Basis.

More often than not, the Warrant is the Claim of a prior argument, and so on ad infinitum. The same is true of the Evidence and all of the parts of an argument. In addition, arguments are key parts of problems—indeed, some of their most important parts—because arguments are used to define what is a problem in the first place and how they ought to be handled in the second place. Even more, they are among the most important parts of Wicked Messes. Indeed, the parts of TAF are generally parts of a Mess.

Most of the time, Warrants are implied rather than stated explicitly. In fact, a great deal of the time they are unconscious. That is, the person making them is not fully aware that they are doing it. In general, they are reflective of a person's entire personal history. In the case of society, they reflect its general history and current conditions.

In the case of tech, a common taken-for-granted Warrant is "The reputation and standing of a technology's developers guarantees that it will work as planned." Or, "The benefits of a proposed technology have been well demonstrated: it's passed every important test with flying colors."

Every argument also has a Backing B. B is the deeper set of underlying assumptions, basic reasons, or values as to why a particular Warrant holds. If the Warrant is not accepted at its face value—which is often the case—then the Backing is necessary in order to support it.

In general, the Backing is the larger set of general Philosophic Assumptions a person has about what is right (Ethics), human nature, the world (Reality itself). Since Backings are generally taken-for-granted, they are likewise mostly implicit and unconscious.

Another way to think of it is as follows. If the Warrant W is the "conceptual bridge" that allows us to go from the Evidence E to the Claim C, then the Backing B is the "foundation" on which the bridge rests.

In the case of tech, the Backing is the credentials of the developers, plus the fact that they are "well-qualified." It's also the general body of stakeholders who will benefit from it. Notice that there is a great deal of overlap among all of the elements of TAF. Again, it's a Mess, and more often than not, a Wicked Mess.

Finally, every argument has a Rebuttal R. In principle, R challenges each and every part of an argument. In terms of the metaphor of arguments as a "bridge" between Evidence and Claims, R attacks E, W, and C as strongly as it can. R thus tries to "tear down the entire bridge and its foundation." For instance, "The Evidence is faulty and/or weak and thus does not support the Claim, etc." Or, "I accept your E, but it leads to a completely opposite C from that which you've claimed."

In the case of the Tech Court, the Rebuttal is the opposition's argument as to why the current proposed technology is not beneficial, indeed, why every part of the so-called supporting argument is deeply flawed. In short, the only reasonable conclusion or Opposing Claim is that the proposed technology should be strictly prohibited.

References

1. Manjo F (2019) Facial recognition must be put on hold. The New York Times, p A29
2. Toulmin S (1958) The uses of argument. Cambridge University Press, New York

The Socially Responsible Tech Company

6

> *[The global information war] is a war without limits and boundaries, and one that we still don't know how to fight. Governments, nonstate actors and terrorists are creating their own narratives that have nothing to do with reality. These false narratives undermine our democracy and the ability of free people to make intelligent choices. The disinfoirmationists [sic] are aided by the big platform companies who benefit as much from the sharing of the false as from the true....Autocrats have learned the same tools once seen as spreading democracy can also undermine it...*
>
> *Stengel* [1]

Proactive Crisis Management (CM) is the backbone of the Socially Responsible Tech Company, SRTC. From its very inception and across its entire lifespan, Proactive CM needs to guide its every action, indeed it's its Raison d'Etre. A SRTC would therefore be constantly on the lookout for the unintended consequences of its technologies, how they can be abused and misused, and how they can be minimized if not prevented altogether. The primary emphasis is on constantly monitoring for harmful effects and ill consequences.

For this reason alone, we review the essential elements of Proactive CM.

Types

The first element of Proactive CM is not only planning, but being prepared for a broad range of different types of crises. Previous research has shown that there are distinct families or types of crises.[1] Thus, there are informational crises as when

---

[1]Mitroff [1].

I. I. Mitroff and R. Storesund, *Techlash*, Management, Change, Strategy and Positive Leadership, https://doi.org/10.1007/978-3-030-43279-9_6

35

unauthorized parties hack into supposedly secure systems and steal highly personal information such as social security, credit, and health data. There are financial crises as in the case where the finances of an institution are mismanaged or subject to criminal malfeasance. There are Public Relations, PR, or reputational crises. For example, Volkswagen caused a huge PR crisis for itself when its concerted efforts to deceive the public with regard to the true values of the pollutants emitted by its cars came to light. Needless to say, Volkswagen's reputation took a severe hit. There are ethical crises as in the case of the Catholic Church where those who are guilty of serious sexual crimes against minors are not brought to the immediate attention of the proper legal authorities. Human Resource or HR crises include deliberate acts of Employee Sabotage. The intentional abuses and misuses of products are contemporary instances of Product Tampering. Natural Disasters are another major category. So are Fires and Explosions. And in today's world, International and Domestic Terrorism are always worrisome possibilities.

These do not by any means exhaust the full range of the possible types of crises for the world is unfortunately continually inventing new ones. Nevertheless, they are sufficient to give a good feel for the different types and the wide range of crises.

Most important of all, while the different types are distinct in that they can be clearly identified and labeled as such, they are not independent. Time and again, it's been shown that *no crisis is ever a single crisis*. Every crisis is potentially both the cause and the effect of every other. A crisis may start in one category or type, but it quickly spreads to all of the others. If one is not prepared, it sets off an uncontrolled chain reaction. Thus, to reiterate, Volkswagen's emission crisis was a HR crisis in that it involved groups of employees up and down the corporate hierarchy. It quickly became a major PR and a financial crisis for the company as whole. In short, all of the various "types" are elements of a Mess!

The key point is that preparing for a limited number of crises is not only inadequate, but actually increases the crisis potential of an individual, organization, society, and in today's world, the entire planet. In brief, it is counterproductive. In fact, it leads to feelings of false security. For this reason, Proactive CM is fundamentally systemic.

All of the aforementioned types apply equally to tech companies. However, in addition, the various types of the Unthinkable are applicable as well. Indeed, they are intertwined. Thus, knowingly and unknowingly, the assumptions that are made about unanticipated stakeholders come back more often than not to haunt one in the form of major crises. Nonetheless, by far, the most dangerous have to do not just with Product Tampering, but with product defects such as latent defects and fatal flaws.

Recognizing and preparing for a broad range of crises ideally leads to setting up specific mechanisms and procedures for the detection of the Early Warning Signals that not only accompany, but precede all crises. Since different types of crises send out different kinds of signals, each crisis calls for its own distinct form of devices and procedures for Signal Detection.

To make CM truly effective, it needs to be an integral part of a company's everyday operations. One of the best ways of doing this is to make it a key part of an already existing program that the organization currently takes seriously such as Quality Control. In fact, Quality Control is a natural home and ally of CM.

Damage Containment and Mitigation

One of the most critical parts of Proactive CM is Damage Containment. And, different types of crises involve different types of Damage Containment Mechanisms and Procedures. What is appropriate for one doesn't necessarily work for others. Thus, a protective barrier designed to keep a fire or oil spill from spreading will not keep a financial crisis from eventually engulfing and destroying an organization. Financial crises necessitate among other things safeguards such as independent auditors and whistleblowers.

One of the most important things about Damage Containment Mechanisms and Procedures is that they cannot be invented in the heat of an actual crisis. BP's 2010 oil spill in the Gulf of Mexico is dramatic testimony to this unfortunate fact of life. Millions of gallons of oil were spilled leading to untold environmental damage and destruction before the errant well was finally capped. Again, the operative words are systemic and proactive.

Ideally, both Signal Detection and Damage Containment direct one to focus on Mitigation. It raises the question as to what one can do to design and redesign one's products and manufacturing processes so that the possibilities of major crises are greatly reduced, if not prevented altogether, and further, that the ultimate goal of ensuring the safety of those who can be potentially harmed is a major priority.

Defense Mechanisms

One of the most critical but unfortunately least discussed factors in the typical accounts of CM are Defense Mechanisms. If Sigmund Freud had done nothing more than discover the existence and functioning of Defense Mechanisms, it would have been more than enough to ensure his lasting fame.

Defense Mechanisms basically exist to protect the minds of individuals from painful events and harmful memories that are too difficult to acknowledge consciously. Thus, if one has been in a war and witnessed the death of a fellow soldier or companion, then one can attempt to shut the entire event out from consciousness through Denial, i.e., by denying that the entire event ever occurred in the first place. The same applies if one has been the subject of a violent attack or incident. Unfortunately, since Denial is never perfect, the repressed event often comes back in the form of nightmares from which the combatants in war typically suffer. Unfortunately, fire and police personnel typically experience this as well.

Early on as one of the key founders of the modern field of CM, Mitroff and his colleagues discovered that there were direct organizational counterparts for every one of the Defense Mechanisms that were originally discovered and which pertained solely to individuals. Thus, in organizations, Denial often takes the form, "We don't have any major problems." With regard to technology, "Our technology will work as intended with no major problems or issues." For another, "Nefarious, malicious actors are not a problem about which we need to worry." "It's a total waste of time and money to Think about the Unthinkable, let alone plan for it." Or, "Despite all the fears and protests, people have always adapted before to new technologies so that there are no serious reasons to believe that we won't do it again. In other words, present concerns are largely overblown."

Disavowal takes the form, "Yes, there are problems, but they are minor. They are not important enough to warrant major attention." Disavowal thereby reduces the complexity, magnitude, and scope of problems and issues to where they can supposedly be downplayed and thereby ignored. Once again, "Fears about technology are greatly overblown."

Projection takes the form, "Someone else is to blame for our problems, not us." In other words, "We're not responsible for problems in any way." In the case of technology, "We're not to blame for those who abuse and misuse our technologies."

Intellectualization is the attitude that "Our technology is perfect." Or, "We can catch all problems before they are too big to handle." For another, "We've thought of everything that could possibly go wrong and planned for it so there's nothing to worry about." As such, it's obviously intertwined with Denial, as are all of the various Defense Mechanisms.

Compartmentalization is the attitude that "A crisis may affect parts of an organization or a technology but not the whole of it." Compartmentalization thus avoids thinking and acting systemically. Crises can thereby be confined as a result.

Grandiosity takes the form, "We are smart, powerful, and competent enough to handle anything!" Again, "We've thought of everything."

Finally, Idealization is the attitude that "Everything will work as planned."

It's been found that the more that an individual or organization subscribes to the Defense Mechanisms outlined above, the more crises they experience. Thus, instead of protecting one, Defense Mechanisms do the exact opposite. Once again, a key element of Thinking the Unthinkable is seriously contemplating how all of one's idealized assumptions can lead to their direct opposite.

Ultra-proactive

If the idea of a Tech Court is a prime example of an _external_ mechanism for helping to ensure that technology will serve us well by causing no harm, then the following is a prime example of a company that did everything that it could _internally_ to help ensure the safety and well-being of its consumers.

Not long after the 1982 Tylenol poisonings, which was the impetus for among other things for the creation of the modern field of CM, Mitroff established The USC Center for Crisis Management, the first academic institution of its kind to study crises of all kinds. The basic purpose of the Center was to do research so that hopefully future crises could be handled better, if not prevented altogether. To further knowledge about CM, the Center had corporate sponsors who not only provided funds for its work, but just as important, gave unfettered access to their organizations so that Mitroff and his colleagues could study how they both prepared for and responded to various types of crises.

In the hopes of learning more about Product Tampering, one of the most important types of crises that apply to all organizations no matter what their business, Mitroff visited a major pharmaceutical company (It's important to stress that with regard to tech, Product Tampering takes the form of the systematic abuse and misuse of technologies.). Mitroff asked the person who agreed to talk with him what his company was doing to combat the ever-present threat of Product Tampering, a major cloud that perpetually hung over the entire Pharma industry. Without losing a beat, he said, "We formed a number of Internal Assassin Teams." To which Mitroff blurted out, "You did what?!"

Yeah, early on we realized that we knew more about our products than anyone else. So one day we held up a bottle of one of our major pain killers and we looked at it as if the cap were the front door and the sides were the walls of a house. We then asked ourselves, 'How could a burglar get in, remain undetected for as long as possible, and thereby do the most damage?'

We quickly learned that there was no way to keep a determined burglar out so that the notion of tamper proof seals was not even a remote possibility. The best we could do was tamper evident seals so that if one of our bottles was breached a consumer would be alerted not to use it.

In the years since, Mitroff has taken a few companies—sadly, all too few—through the exercise of an Internal Assassin Team as it applied to their organization. It allowed them to imagine and thus confront the all-too-real possibility of the worse, the Unthinkable, happening to them and their company. The point is that one needs explicit permission to imagine the worse and then to do everything possible to prevent it. Thinking the Unthinkable does not happen naturally on its own.

Risk Management RM

If only for the reason that RM is often confused with CM, it's imperative that we say a few words about it. The two are definitively not the same.

RM is based on the concept of Expected Value or EV, which derives from the theory of Probability. To take an overly simple example, if we have an unbiased coin, one for which the probability of getting a heads is ½ and is the same for a tails, then if we throw a coin 100 times, we would expect to get 50 heads and 50 tails. Suppose every time we get a heads we win $1 and every time we get a tails we lose

$2. Then, we would expect on average to get 50x \$1–50x \$2 or to lose \$50 in 100 tosses. The formula is EV = The Probability of Event One x The Payoff or Loss of Event One + The Probability of Event Two x The Payoff or Loss of Event Two, where there can of course be more than just two events and outcomes.

There are at least four possibilities with this supposedly simple procedure:

1. We know both the exact probabilities and outcomes or consequences of an event;
2. We know the probabilities but not the consequences;
3. We don't know the probabilities but we know the consequences.
4. We know neither the probabilities nor the consequences.

Case 1 is the prototype of an exercise where everything is supposedly known or given. As we've indicated, this is rarely the case. Indeed, it really happens only in textbooks.

Case 2 is best exemplified by wildfires. While the probabilities are not known exactly, wildfires are nearly certain to occur in every fire season in the Western States. It would be the extremely rare season in which there was not at least one major fire. And, while the consequences from previous years are known, they vary dramatically from year to year.

Case 3 typifies earthquakes where the consequences are known to be high, but when and where a major earthquake will occur is "problematic."

Case 4 is that of Wicked Messes. It also typifies "Unknown Unknowns." That is, where in cases 2 and 3, We Know What Don't Know, in case 4, We Don't Know What We Don't Know.

Unfortunately, RM contains even greater defects. First of all, it's rarely systemic in that typically one doesn't consider how one risk can set off a chain reaction of others as is the case with CM. Second, it does not take into account Defense Mechanisms, which prevent far too many organizations from engaging in RM and CM in the first place.

Also, RM does not take into account that we don't experience the Expected Value of an event but the actual costs. Thus, while the probability of a \$1,000,000 house burning down in a particular area may be low, say 1 in a 1000, the replacement cost is not 0.001 times \$1,000,000 or \$1000 but more likely \$1,200,000 and up.

For these and other reasons, we are not champions of RM. If RM is done at all, we insist that it be part of a broader, concerted effort in CM.

One of the saddest aspects of the whole issue of Crisis Preparedness is the fact that at best only 10–15% of companies have anything approaching a viable, let alone ideal, program in CM. This is in spite of the fact that those companies that are Crisis Prepared not only experience substantially fewer crises, but are significantly more profitable. If ever there was testimony to the power of Defense Mechanisms, namely "We don't have any problems for which we need to prepare," etc., this is it!

The Need for Systemic Thinking

Finally, we want to conclude this chapter by emphasizing once again the systemic nature of CM. The best way to do this is by understanding the kind of Inquiry System (IS) that is the foundation of Proactive CM.

The History of Philosophy recognizes at least five different and distinct methods of Inquiry (Inquiry Systems or IS's) for gaining knowledge if not the Truth[2]:

1. Expert Consensus or Empiricism;
2. Analytic Modeling or Rationalism;
3. Multiple Perspectives;
4. Conflictual Modeling of Dialectics; and
5. Pragmatist or Systems Thinking.

The first model or IS is predicated on the agreement between different experts. That is, the tighter the agreement between the assessments of set of independent, recognized experts with regard to the "Facts" of a situation, the more that the "Facts" are the Truth. The latest incarnation of this IS is "Big Data" which forms the basis of algorithms which are in turn the basis of AI programs.

The second IS asserts the primacy—indeed, superiority—of Single Analytic Models over data no matter how big the datasets are. In this way, the first two IS's are best suited to exercises where problems and issues are well-defined.

The third IS is based on the strong presumption that no single group of experts or model is ever sufficient to capture a situation fully. The more complex and critical the situation, the more we need to see how it appears to different stakeholders—indeed how they "represent it." At the least, we need to negotiate the different perspectives of different experts. Thus, to take an important case, we wouldn't expect a social worker, a psychologist, a medical doctor, parent, etc. to have the same view of drug addiction, let alone how to treat it successfully. It's not that one perspective or viewpoint is right and the others are wrong, but that all of them are merely picking up "One Aspect Of A Complex Truth."

The fourth IS says that we need to have the strongest debate we can muster before we decide a critical situation, as in the case of the Airman in Alaska who had a debate with himself as to whether the "large object coming over the horizon" was actually a swarm of enemy missiles or not.

As we have stressed repeatedly, all of the issues connected with tech deserve the strongest debate to which we can subject them. To take a single case, with regard to the Claim that Social Media can and should speed up conversations—more importantly, decision-making—the counterargument is that to further deliberation and reflection, Social Media should be slowed down. If not, it does not necessarily lead to positive outcomes.

---

[2]See Mitroff and Linstone [2].

Finally, the last IS is the foundation for Systems Thinking. It insists that all problems of any importance are parts of Wicked Messes and must be understood as such. Accordingly, it's the foundation of the thinking and behavior of the Socially Responsible Tech Company.

References

1. Mitroff II (2005) Why some companies emerge stronger and Bteer from a crisis: 7 essential lessons for surviving disaster. Amacom, New York
2. Mitroff II, Linstone HA (1993) The unbounded mind, breaking the chains of traditional business thinking. Oxford University Press, New York

The Moral Underpinnings of Technology

7

> *...The deeper problem is the overwhelming concentration of technical, financial and moral power in the hands of people who lack the training, experience, wisdom, trustworthiness, humility, and incentives to exercise that power responsibly.*
>
> *Stephens* [1]

As we've stressed throughout, Ethics is one of the prime features of Wicked Messes. Indeed, it's one of the most critical elements of Pragmatist Inquiry Systems. It certainly is one of the most important components of all crises, indeed of everything that humans do. As such, it plays a major role in Thinking the Unthinkable. For this reason alone, we need to address the Ethical and Moral underpinnings of technology.

First of all, it's often asked, "How does Ethics play a part in Natural Disasters that by definition are not caused by humans?" The short answer is that it is we not Mother Nature who make the crucial decisions where and to what standards to build houses and other structures. As a result, all Disasters bear the indelible imprint of humans, and in this sense, are Human-Caused.

Utilitarianism and Deontology are two of the major schools and prominent theories of Ethics. In its original and classic formulation, the basic idea of Utilitarianism is "the greatest good for the greatest number." In the case of technology, something is "Ethical" if its benefits clearly exceed its disbenefits. The customary objection is, "What counts as benefits for which stakeholders, and why not for others?" And of course, who sets the "Ethical Threshold" as it were above which we say something is "Ethical or not"? That is, what the benefits are and how and why they exceed the disbenefits.

When it comes to technology, Utilitarianism fares extremely poor. If the proponents of a technology overly exaggerate and hype its benefits and downplay, if

© The Author(s) 2020
I. I. Mitroff and R. Storesund, *Techlash*, Management, Change,
Strategy and Positive Leadership, https://doi.org/10.1007/978-3-030-43279-9_7

not ignore altogether, the costs or disbenefits, then Utilitarianism fails miserably as an adequate base for assessing the Ethical standing of a technology.

The basic idea of Deontology is that what is Ethical should not be based on consequences per se, but on whether a formal rule or principle can be "willed without contradiction as a general principle that is applicable to, and thereby Binding on, all of humanity at all times and places." According to the great German Philosopher Immanuel Kant, who is generally credited as the "Father" of Deontology, the classic case is that of lying. One cannot will lying as a general rule for it is inherently contradictory and thereby self-defeating. It would defeat our ever believing in anyone, and hence, The Truth! In the case of technology, one of the primary forms of Deontology is that "There should be as few restraints as possible placed upon innovation for it's absolutely essential to human progress; therefore, it must be protected at all costs."[1] Again, the primary question is "essential to whom, and in all cases at all costs?"

Given the despicable behavior of today's tech companies, Kant must be rolling in his grave. We believe that he would be absolutely horrified by the actions of Facebook's CEO Mark Zuckerberg's misguided notion of "free expression," or whatever he calls it, to allow political ads that contain outright falsehoods and lies. The fact that there exists a deliberate mechanism to spread lies worldwide at the click of a mouse is nothing less than Kant's worst nightmare come to life. It's the height of Unethical behavior.

(It should be noted that YouTube fares no better when it comes to carrying political ads that contain outright lies. The point is that Unethical behavior is unfortunately not confined to one tech company and one alone.)

Such behavior has not gone without significant pushback. A total of 250 Facebook's employees have dissented in the strongest possible terms. In an open letter to Zuckerberg, they write:

> Free speech and paid speech are not the same.
>
> Misinformation affects us all. Our current policies on fact-checking people in political office, or those running for office, are a threat to what FB stands for. We strongly object to this policy as it stands. It doesn't protect voices, but instead allows politicians to weaponize our platform by targeting people who believe that content posted by political figures is trustworthy.[2]

In spite of the historic importance of Utilitarianism and Deontology, we want to take a different approach. A particular method known as Moral Foundations Theory or MFT is especially relevant for assessing the Moral and Ethical issues associated with technology.

---

[1]See Mitroff [2].
[2]Issac [3].

MFT

MFT derives from the fields of evolutionary psychology and cultural anthropology.[3] Compared to Utilitarianism and Deontology, MFT is based on how people actually experience Ethics and Morality. According to MFT, a person's Moral intuitions or instincts (feelings) not only come first, but take precedence over reason, and thus override it. In essence, because they are "too abstract," Utilitarianism and Deontology do not figure importantly in our everyday lives. Reason not only plays a secondary role, but basically justifies a person's initial moral instincts.

MFT employs a powerful metaphor to make its key point about the role of reason. The mind is likened to a large elephant with a tiny rider that attempts to control the huge beast. The "rider" is reason, and the "elephant" is all the nonrational emotions and instincts that govern human behavior. Over time, the rider does indeed gain control, but only by recognizing and making peace with the elephant.

MFT is based on two basic principles: Individualizing and Binding. The Individualizing Principles (care and fairness) serve to protect individuals from harm and injustice. The Binding Principles (loyalty, authority, and sanctity) help to establish groups that are loyal, cohesive, and cooperative. In this way, it also serves to protect them as well. To elaborate:

1. Care: Evolution endowed us with the ability to form deep attachments with one another. It allows us to be able to commiserate with the pain and distress of others. It's responsible for the fundamental virtues of kindness, gentleness, nurturance, sympathy, and compassion. It's the underlying basis of the principle "Do no harm!", which is of course an expression of a Deontological rule, but in a form to which one can relate more directly. To say the least, tech comes up short because care is not necessarily the basic, underlying principle that guides the behavior and goal of tech companies. It's certainly not seen as central to their basic mission.

2. Fairness: Survival necessitated that our ancestors not only had to learn to work well with their immediate families and clans, but also with others with whom they were not directly related. In brief, reciprocal altruism is the basis of the fundamental ideas of justice, rights, autonomy, and fairness. Tech also come up short here as well because if it embraced the principle of fairness then it would automatically include a much broader set of stakeholders such as a company's surrounding communities, the environment, and the most distant set of all, future generations.

3. Loyalty: Fidelity to a particular tribe was also necessary for survival. It's the basis of the principle of self-sacrifice for the group as a whole. It often finds expression in the desire "to be of service to a cause 'greater than oneself.'" If anything, tech is overly loyal to a narrow set of stakeholders, e.g., Venture Capitalists being among the primary who provide the basic operating capital to start a company.

---

[3]Graham et al. [4, 5].

4. Authority: As primates, our ancestors had to learn to function and thereby survive in groups that contained hierarchies. The recognition of authority is necessary both for the ability to lead and to follow. It also plays a crucial role in the ability and willingness to honor traditions. The sad fact of the matter is that very few organizations give their members the power to challenge authority.
5. Sanctity: Avoiding contact with contaminants was and is crucial to survival. Evolution endowed our ancestors with the ability to be disgusted by and avoid what was found to be "dirty" and "unclean." Sanctity is the basis of traditions that emphasize the moral significance of maintaining the purity of the body. Purity takes on a particular role in an appalling number of organizations, namely they see themselves and their motives as "pure."

The Ant-Vaccine Movement

The Ant-Vaccine Movement provides a very interesting and relevant case. In 2017, using MFT, researchers surveyed 1007 American parents.[4] "…Those most resistant to vaccines scored highest in two values: purity ('my body is a temple') and liberty ('I want to make my child's health care decisions')". "Deference to authority" also figured importantly in regard to whether one was willing to "adhere to the advice of experts like a pediatrician or the C.D.C."

Proactive CM

MFT serves as the underlying Ethical basis of Proactive CM organizations. They are built on the two Individualizing Principles of care and fairness. The three Binding Principles of Loyalty, Authority, and Sanctity further reinforce them. In fact, the validity of the Binding Principles depends on whether they support the Individualizing Principles.

The Jungian Framework

A special framework due to the pioneering work of the highly influential Swiss psychoanalyst Carl Jung is also extremely relevant for the topic at hand.[5] It sheds needed additional light on the difficult issue of Ethics.

As a highly educated European early in the twentieth century, Jung was well versed in history, literature, philosophy, and psychology, to mention only a few of the many subjects with which he was familiar. No matter what the individual

---

[4]Hoffma [6].
[5]See Ian I. Mitroff, Lindan B. Hill, and Can M. Alpaslan, Op. Cit.

subject that he examined, time and again he observed the same differences in perspectives with regard to how people approached the topic at hand.

At the one of the spectrum were those whom he termed Sensing Types. They not only focused on Details, Specifics, and the Present, but even more fundamental, Details, etc., were the underlying basis of Reality, and as a result, the only legitimate way of obtaining knowledge. One didn't know something for certain unless one had hard, precise, and verifiable facts about the specific matter at hand.

At the other end were Intuitive Types whose focus was on the Bigger Picture, Systems, and Wholes. They also were tuned to hypotheticals and the future. In other words, they didn't shy away from What-Ifs. Indeed, they embraced them for they were the key to The New.

For Jung, it was absolutely not the case where one type or perspective was "right" and the other was "wrong." Instead, they both "saw" and thereby "constructed" the world differently. Not only was it vital to understand each of them, but not to put either of them down. Far more than they were able to acknowledge, they not only needed one another to keep them honest, but to "round out" their perspectives, and thus give a "more complete picture of the world."

The differences between Sensing and Intuitive Types constituted one dimension of the Jungian Framework. The other, which was independent and thereby "orthogonal" to the first, concerned the differences between thinking and feeling types. The focus of thinking types was on impersonal modes of analysis such as economics, logic, mathematics, and hard science. In sharp contrast, the focus of feeling types was on people, i.e., the Human Impacts of ideas, plans, policies, in other words, Highly Personal Ways of Knowing. It's important to note that by feeling, Jung did not mean emotional, for each and every one of the psychological types could be extremely "emotional' in defending its view of Reality.

Putting the two dimensions together results in four very different personality types: (1) Sensing-Thinking or ST's for short; (2) Intuitive-Thinking or NT's as they are referred to; (3) Intuitive-Feeling or NF's; and (4) Sensing-Feeling or SF's (N is used for intuition because the letter I is reserved for introverts. Conversely, E stands for extroverts, which is another key dimension of the Jungian Framework.). These four have very different and distinct approaches to technology, if not to the world and Reality in general.

ST's are focused exclusively on the known, verifiable properties and benefits of today's technologies. In effect, they are not only comfortable with reducing everything to exercises, but insist on it as a primary condition before knowledge can be said to result. In this way, ST is inherently evolutionary when it comes to technology. Not only does it extend the Present, but it doesn't threaten to disrupt it. At its best, it doesn't release a new technology until it is absolutely assured that it will perform as intended, and thereby cause no harm, or as little as possible. It's great weakness is its inability and unwillingness to envision how unknown, totally new and unexpected nefarious actors will exploit a technology's underlying weaknesses and inherent defects for their benefit. At its worst, not only is it unconcerned with unintended consequences, but is oblivious to them.

In contrast, NT's are primarily concerned with groundbreaking, disruptive technologies that no one has ever imagined, let alone seen before. Thus, where ST's are Evolutionary, NT's are Revolutionary. Their great strength is that they are able to envision and bring about the future. Nonetheless, for all their ability to think and act systemically, their greatest weakness is also their disregard for how nefarious actors will take advantage of their marvelous creations. They are also unconcerned with the unintended consequences of their inventions and actions.

NF's are primarily concerned with the Human Impacts of technology, not only on the Present and society in general, but on future generations. This is their greatest strength. Their weakness is their general suspicion of technology period, not to mention their disdain for technologists because of their general lack of concern with people.

In our unremitting criticism of technology and especially the narrow thinking of far too many technologists, we are the first to admit that this book smacks heavily of NF. However, it's not devoid of the other perspectives as well.

SF's are primarily concerned with how technology affects them and their immediate friends and families. Unless something can be shown to have concrete and distinct social benefits to them personally, they are generally opposed to it.

It should be clear that these four have very different approaches to Ethics. Thus, ST's give their primary allegiance to Cost/Benefit Analysis. That is, a technology is warranted if the benefits clearly exceed the costs. In contrast, NT's also use Cost/Benefit Analysis but with a completely different twist. They run multiple Cost/Benefit Analyses to make sure they have not overlooked anything serious. Thus, where ST is partial to the first two IS's—Expert Consensus and Analytic Modeling—NT's are partial to Multiple Perspective IS's.

Given that Feeling overrides Thinking, NF's are in effect practitioners of MFT. Harm is extended to all of humanity. SF's practice MFT as well except they do it with a focus exclusively on their immediate families and friends.

Ideally, these four types need to work together to overcome their inherent weaknesses and limitations. Indeed, all of them are needed to give a systemic overview. Sadly, because of our work with countless groups over the years, the authors have seen repeatedly how difficult it is for the different types just to listen to one another, let alone work together. They literally speak and live in different realities. For this reason, where we can, we've exposed them to the Jungian Framework so that they can work more productively with one another.

The point is that our approaches to technology, the Unthinkable itself, are shaped by fundamental differences in psychology. These can either be an aid to how we think about and manage technology, or the greatest obstacle. In the words of Pogo, "We've met the enemy and he is us!"

Concluding Remarks

We are also the first to acknowledge that MFT is an ideal (As we've used it, the Jungian Framework is as well.). Nonetheless, its "primary virtue" is that it gives an explicit scorecard against which to measure the CM behavior of every organization. Sadly, tech does not fare well with regard to each of the core principles.

References

1. Stephens B (2019) Facebook's unintended consequence. The New York Times, Saturday, 4 May 2019, p A19
2. Mitroff II (2018) Technology run amok: crisis management in the digital age. Palgrave Macmillan, New York
3. Issac M (2019) Dissent erupts at Facebook. The New York Times, Tuesday, 29 Oct 2019, p B1
4. Graham J, Nosek BA, Haidt J, Iyer R, Koleva S, Ditto PH (2011) Mapping the moral domain. J Pers Soc Psychol 101(2):366
5. Graham J, Haidt J, Koleva S, Motyl M, Iyer R, Wojcik SP, Ditto PH (2013) Moral foundations theory: the pragmatic validity of moral pluralism. In: Advances in experimental social psychology, vol 47. Academic Press, Cambridge, pp 55–130
6. Hoffma J (2019) How Ant-Vaccine sentiment took hold in US. The New York Times, Tuesday, 24 Sept 2019, p A21

Epilogue the Future of Technology

The Facebook Page 'Stop Mandatory Vaccination' has more than 140,000 followers. Its moderators regularly post material that is framed to serve as evidence for this community that vaccines are harmful or ineffective, including news stories, scientific papers and interviews with prominent vaccine skeptics. On other Facebook group pages, thousands of concerned parents ask and answer questions about vaccine safety, often sharing scientific papers and legal advice supporting antivaccination [sic] efforts. Participants in these online communities care very much about whether vaccines are harmful and actively try to learn the truth. Yet they come to dangerously wrong conclusions. How does this happen?

O'Connor and Weatherall [1]

Major Themes

In a recent *New York Times* interview, Brad Smith–Microsoft's President and Silicon Valley's self-appointed person to assuage the public's growing apprehensions about tech—freely admitted that "Until you acknowledge the problems, you can't solve them. And nobody is going to believe you're trying."[1]

While well intended, if readers hoped to find anything approaching a meaningful and comprehensive discussion of tech's many problems, they'd look in vain. The following is not only a summary of the key themes of the book, but the major problems facing tech:

1. *"Who and what is, and will be, human?"* This is arguably the most worrisome of all of the issues. As we've maintained, it's part of our relentless desire to direct and thereby be the masters of the next stages of human evolution, if not the ultimate desire to exert total control over nature. Accordingly, we've developed the godlike abilities to make changes to our genetic code. The all-too-real fear that we are on the road to producing to "Designer Humans." It's made worse by the fact that we don't have the necessary understanding of what we are truly

---

[1]**On Stage**, "Microsoft's President on Silicon Valley in Crisis," *The New York Times*, Thursday, 19, 2019, p. F3.

© The Editor(s) (if applicable) and The Author(s) 2020
I. I. Mitroff and R. Storesund, *Techlash*, Management, Change,
Strategy and Positive Leadership, https://doi.org/10.1007/978-3-030-43279-9

messing with, and thus whether we'll be able to manage it. It's further aggravated by those who want to put AI chips directly in our brains so that we'll not only be able to communicate seamlessly with all of our marvelous gadgets, but be "new and improved human beings." As a result, the overriding question is, "Who and what will be human?" To our great misfortune, if not demise, our godlike powers are not accompanied by the godlike wisdom to know when not to do something just because we can.

2. *"Who and what are real?"* AI allows for the production of videos of people doing and saying things that they never would but look so "real" such that they fool even experts as to their authenticity. The fundamental nature of Reality and who and what one can trust are called into question as never before. How can any society—indeed societies worldwide—survive repeated assaults on whom and what are real?

3. *"Who can I truly trust to protect my confidential and most personal information?"* Tech companies now collect as much and in many cases more information about us than the government. Worse, the business models of Facebook and the like are tied directly into selling that information to unscrupulous third parties for their immediate gain and profit. Whom can one really trust to protect one's vital information? Nothing is as seriously injurious to societies worldwide as the erosion of trust.

4. *"Whom and what can I trust to be truthful?"* Social Media have become principal platforms for the production and dissemination of conspiracy theories, dis- and misinformation, scandalous rumors, etc. The fundamental nature of Truth has been seriously jeopardized, if not damaged irreparably. Further, given the unmitigated production of hate speech and our pitiful attempts to curtail it, Civility has also been seriously wounded as well. If we had intentionally set out to create deliberate instruments for the wholesale destruction of Civility and Truth, we couldn't have done a better job than that of so-called Social Media. They're better branded for what they really are, "Anti-Social Media!"

5. *"What if anything is being done to protect us from the unintended consequences and the abuses and misuses of technology?"* As a general rule, technologists are so enthralled by their wondrous creations that not only are they unable, but unwilling to consider how they can and will be systematically exploited by nefarious actors to their advantage. They overly hype the positive benefits of their magical devices to the near, if not total, exclusion of serious consideration of anything negative. Social Media are the premier examples. What was supposed to "connect us" and thus "bring us closer together" is now one of the greatest forces responsible for driving us further apart. Not only have they served as relentless platforms for Cyber Bullying, but they have allowed foreign governments to interfere in our elections. In addition, a growing body of evidence confirms that the more young people, and even adults, use Social Media, the more isolated, lonelier, and depressed they are. The very thing that was supposed to raise our self-esteem is directly responsible for lowering it. They hinder our both ability and desire to Thinking the Unthinkable.

In no way do these exhaust all of the problems connected with technology, but they are sufficient to illustrate the many troublesome concerns they have spawned.

The short of it is that tech needs to be strictly regulated. As we discussed, a new federal agency, The Office for Socially Responsible Tech needs to be created and scrupulously maintained.

No society can withstand continuous assaults on Truth, Reality, and Trust. They are the very foundations of society. As such, they must be protected by appropriate rules and regulations if need be.

The Way Forward

One of the key messages of this book is that unless tech companies become socially responsible, the backlash against them will only get worse, thereby preventing us from reaping the many benefits of technology. If we are to continue to receive its many blessings, something fundamental needs to change. Principally, we need to change our basic attitudes toward the meaning of knowledge—Truth itself—and how to best obtain it. Our traditional means of producing knowledge are outdated. They are not up to the task of coping with a complex world of Wicked Messes.

We are under the grip of methods that were appropriate for simpler times—in essence, a world of exercises. To recap, in its original formulation, Empiricism no longer applies. Reality is not a set of "hard facts" with which everyone agrees. It's not just that different stakeholders have different facts, which they do, but that they interpret the "same facts" differently because of their differing perspectives, experiences, and underlying values. Ever since the great philosopher Immanuel Kant—indeed, it's one of his greatest contributions—we've known that "facts" are not independent of the theories and values that are necessary to unearth them. It requires an underlying theory of the phenomenon in which we are "interested" (hence, our values are involved from the very beginning) to collect the "facts" that are pertinent to our investigations. Any old set of facts or data will not do. Furthermore, when it comes to "scientific facts," it requires a great deal of training to produce them.

This does not mean that there are no "hard facts" whatsoever. We accept it as a "well-established fact" that 97% of "reputable climate scientists"—over 11,000 in the most recent count—are in strong agreement that humans are the major source of Global Warming, and as a result, the Earth is getting dangerously warmer. But then, they are in substantial agreement because of their shared backgrounds in science and independent assessments of the "facts" regarding Global Warming. The point is that facts don't just "announce themselves" independently of our backgrounds, interests, training, and theories regarding them.

In sum, as it's been traditionally formulated, classical Empiricism is not an adequate base for the formulation of algorithms for critical decisions. But then neither is classic Rationalism.

If hard facts are the basic building blocks of Empirical knowledge, then incontrovertible propositions are the fundamental building blocks of Rationalism. Historically, they are ideas and propositions that are taken to be so "certain" such that they cannot be doubted. The so-called fundamental laws of Classical Logic on which all rational thought and reasoning itself depend are a prominent case in point. Given that one needs to presuppose some ideas or theories that one takes for granted in order to function, Rationalism always plays a central role in the construction of knowledge. Nonetheless, what is necessarily true beyond all doubt for one set of stakeholders at one point in time is not necessarily true for others. For this reason, different theories of the issues at hand are among the prime components of every Wicked Mess.

In our earlier discussion of Inquiry Systems, we not only referred to Dialectics upon which this book has drawn heavily, but we also referred to Pragmatist IS's as the embodiment of Systems Thinking. The essence of Pragmatism is captured in the following: "Truth is that which Makes an Ethical (some would say Spiritual) Difference in the Quality of one's life." Thus, Epistemology—the nature of "Truth" and how best to obtain it—Ethics—the nature of the "Good"—and Aesthetics— what is harmonious and pleasing, are inseparable, where the term Quality is a synonym for Aesthetics, are all fundamentally related. Indeed, a fundamental tenant of Pragmatism is that they are not merely interdependent, but inseparable. The determination of what is true is not independent of what is Ethical, and how it's presented, and thus Aesthetics.

The little word "Makes" is highly critical for according to Pragmatism, one does not have Truth until one undertakes a series of Ethical Actions designed to right a wrong and thereby improve one's condition. Further, one does not have Truth independently of the Ethical Implementation of one's ideally Ethical ideas. In other words, "Truth" is not merely a set of ideas in a paper or book, however, good they may be! It consists of Ethical Actions.

But by far, the most important contribution of modern Pragmatism is the concept of Wicked Messes. It not only recognizes, but extends the complex nature of Reality far beyond anything that's come before it.

Pragmatism also stresses that if Technology magnifies and extends the senses, then ideally the Humanities magnify and extend the sensibilities. They inform us about what's worth doing, and hopefully, how to guard against what is not. In this regard, Pragmatism stresses the inseparable bond between the Sciences and Humanities. Much of the trouble in which technology finds itself is due to the fact that it has severed its relationship with the Humanities, and as a result, failed to develop it as needed for today's world.

That we are a high-tech society is nothing but an obvious truism, if not a cliché. We are becoming more dependent upon and infused with technology every day. The supreme challenge is to make and to manage those technologies that serve and elevate the human spirit, not demean and subjugate it.

As we've argued repeatedly, the key to managing technology for the greater good is Thinking the Unthinkable and then doing everything we can to thwart it. If we have accomplished anything, we hope we've shown that not only must it be done, but that it can be.

We can think of no more fitting words on which to end than those of the esteemed author Aldous Huxley:

> Our business is to be aware of what is happening, and then to use our imagination to see what might happen, how this might be abused, and then, if possible, to see that the enormous powers which we now possess thanks to these scientific and technological advances be used for the benefit of human beings and not for their denigration.[2]

Reference

1. O'Connor C, Weatherall JO (2019) Why we trust lies. Scientific American, Sept 2019, p 36.

---

[2]Aldous Huxley, cited in Editor's Note, http://CaliforniaMag.Org, Winter 2019, Vol. 130, No. 4, P6.

Printed by Printforce, United Kingdom